Science Learning in the Early Years

ACTIVITIES FOR PreK-2

Science Learning in the Early Years

ACTIVITIES FOR PreK–2

PEGGY ASHBROOK

press

National Science Teachers Association

Arlington, Virginia

National Science Teachers Association

Claire Reinburg, Director
Wendy Rubin, Managing Editor
Rachel Ledbetter, Associate Editor
Amanda O'Brien, Associate Editor
Donna Yudkin, Book Acquisitions Coordinator

ART AND DESIGN
Will Thomas Jr., Director
Joe Butera, Senior Graphic Designer, cover and
  interior design

PRINTING AND PRODUCTION
Catherine Lorrain, Director

NATIONAL SCIENCE TEACHERS ASSOCIATION
David L. Evans, Executive Director
David Beacom, Publisher

1840 Wilson Blvd., Arlington, VA 22201
www.nsta.org/store
For customer service inquiries, please call 800-277-5300.

Cataloging-in-Publication Data for this book are available from the Library of Congress.

ISBN: 978-1-941316-33-7
e-ISBN: 978-1-941316-38-2

Contents

Part III. *The Early Years Column and Companion Blog*

Acknowledgments

The children who so joyfully explored science concepts with me must be thanked first. The programs that sought to include science and engineering education in their curriculum, and the parents who support such programs, made this work possible.

The Early Years columns written for *Science and Children* owe their clarity to the excellent editors who supported my growth as a writer: Linda Froschauer, Valynda Mayes, Stephanie Morrow, Stefanie Simmons, and Monica Zerry. Their understanding of science and engineering concepts and education practices, combined with their wordsmith genius and gentle tutoring, make the work worthy to publish as a book. Copyeditor Pat Freedman's patient corrections of style and grammar, checking and cross-checking details, and keen sense of order brought clarity to the book. Her cheerful requests made the revising enjoyable. However, any errors or omissions are mine.

I owe much of my understanding to colleagues in early childhood education who taught me to honor children's capabilities by going beyond "an activity": Karen Worth, Ingrid Chalufour, Cindy Hoisington, and Jeff Winokur at Education Development Center, Inc., and those at the University of Northern Iowa Regents' Center: Beth Van Meeteren, Rosemary Geiken, Sonia Yoshizawa, and Betty Zan.

The colleagues who provided feedback when I turned to them for expertise improved the work: Ken Roy, Marie Faust Evitt, Katina Kearney, and Whitney White.

Thank you to the educators who provided the lists of resources for the original "Teacher's Picks": Fred Arnold, Charlene K. Dindo, Marie Faust Evitt, Yvonne Fogelman, Sarah Glassco, Mary Ann Hoffman, Sarah Pounders, Juliana Texley, and Nancy Tooker.

Thank you to NSTA Press editors Claire Reinburg and Wendy Rubin, and Donna Yudkin, Book Acquisitions Coordinator, for publishing resources for early childhood educators. Thank you to the NSTA Press reviewers and other readers—I appreciate your thoughtful comments because they helped me refine the ideas and reduce confusing statements.

I have benefited from the knowledge and support of the communities of colleagues in the National Association for the Education of Young Children Early Childhood Science Interest Forum and the National Science Teachers Association Early Childhood Community Forum—thank you!

I am grateful to my husband, Darryl François, for our shared interest in science and for his support of my writing.

Introduction

The past few years have seen an increase in understanding about the importance of providing young children with the opportunity to explore their world and confront challenging concepts in science, engineering, and technology. This has led to an increasing number of resources for teachers of young children, but few of these resources are designed to support teachers both practically and theoretically—to provide them with both classroom ideas and an understanding of how to use them effectively. *Science Learning in the Early Years: Activities for PreK–2* does just that. Thus, it is a pleasure to write this introduction.

Through her work with the National Science Teachers Association and the National Association for the Education of Young Children, Peggy Ashbrook has been a driving force in bringing an understanding of the importance of high-quality science and engineering teaching and learning in the early years not only to the early childhood community but also to the science education community writ large. She has also been a tireless writer. Her column, *The Early Years,* in the journal *Science and Children;* the column's companion blog, *Early Years;* and the book *Science Is Simple* (Ashbrook 2003) provide a wide range of ideas for bringing these subjects into the classroom, helping teachers, parents, and other caregivers see what children engaged in science and engineering really look like and how these subjects fit into the life of a classroom or other child-care setting.

This book brings together in one place many entries from Peggy's column and comments and ideas from her blog. Peggy sets the context for these activities in Part I and returns to it throughout the book. Herein lies one of the strengths of the book. In Part I she makes clear that the work she has done has a direct link to standards and frameworks. She explains that these experiences have the capacity to build a foundation for children's later understanding of the core ideas, crosscutting concepts, and science and engineering practices outlined in the *Next Generation Science Standards* (NGSS Lead States 2013). But she also makes clear that for the youngest children, science and engineering experiences should be expansive, and not limited to the specific performance expectations for kindergarten and elementary grades. They should be play-based delightful experiences that enhance and maintain children's natural curiosity and abilities. She also is careful to suggest how science and engineering can support other goals, objectives, and standards, whether they concern overall cognitive development, the development of literacy and numeracy skills, social and emotional competence, or physical development.

A teacher herself, Peggy addresses this book to teachers and others who work with young children. Critical to the tone and content is the fact that she writes from her own years of experience in the classroom and a deep practical as well as theoretical understanding of teaching and learning science and engineering and of young children and their teachers. As you read the column entries and related blog posts in Part III, you feel you are talking with a fellow practitioner. She's been there and done these things and has learned and wants to share. And the blog posts add another important perspective—that we all have something to contribute and that it is through conversation and collaboration that we learn.

But this is not just another book of activities to do week after week; in its structure it is a resource to use to put together an interesting range of activities to enrich a teacher's carefully planned focus on a limited number of concepts. And this is another of its strengths. Peggy reminds the reader that science and engineering learning is not about isolated fun activities but rather is about exploration, investigation, and reflection over time focused on foundational concepts and the use of science and engineering practices. It is exciting, it is challenging, and it is fun to engage teachers and children alike. Peggy makes a careful distinction between learning facts and bits of information and constructing an understanding of basic concepts. She insists over and over that effective inquiry-based science and engineering teaching and learning can only happen when learning is focused on important ideas and children are given the opportunity to explore ideas in depth and follow interesting questions guided by a well-prepared and curious adult. Thus, she says in Chapter 1 that this book is a guide to activities that should be part of a larger science inquiry, "just as a side dish is part of a meal that itself is part of a day's nutrition." And she reminds readers that these activities are not stand-alone science or engineering curricula but rather are small steps in a journey of science inquiry.

This is a book that adults working with children in many different settings will want to keep on hand and return to as a resource and guide for planning and implementing rich science and engineering experiences for children.

Karen Worth
Instructor
Department of Elementary and Special Education
Wheelock College

References

Ashbrook, P. 2003. *Science is simple: Over 250 activities for preschoolers.* Beltsville, MD: Gryphon House.

NGSS Lead States. 2013. *Next Generation Science Standards: For states, by states.* Washington, DC: National Academies Press. *www.nextgenscience.org/next-generation-science-standards.*

A Word About Safety

Young children repurpose materials in ways that early childhood educators may not foresee. For example, instead of feeling the texture of salt in a cup while preparing to make playdough, children may unexpectedly blow into the cup, sending salt grains into their eyes. By reading about safe practices described in the *Safety First* column in *Science and Children* (Roy 2012–2015—see list of column entries at the end of this section), we can learn from the experiences of others. We can educate ourselves about how to protect children from accidents, thus allowing for a safer teaching and learning experience.

Find the online Safety Data Sheets (SDS) for common classroom materials, such as paint, and familiarize yourself with needed precautions. For example, when using tempera paints under "normal conditions" I do not require children to use safety goggles based on SDS information. However, when we test the paint's ability to splash while making a painting in the style of Jackson Pollock, I have the children protect their eyes with indirectly vented chemical-splash goggles.

In addition to the safety precautions we use on a daily basis, such as covering outlets and removing choking hazards for children who put objects in their mouths, we need to protect children from hazards they cannot foresee, such as germs on unwashed hands.

Each column entry in this book has alerts (signaled by the word *CAUTION*) for safer practices. These safer practices should be used in addition to your own judgment and licensing safety requirements based on legal safety standards and better professional practices. Before any activity or investigation is done, always review important safety information with students and volunteers.

Disclaimer: The safety precautions of each activity are based in part on use of the recommended materials and instructions, legal safety standards, and better professional practices. Selection of alternative materials or procedures for these activities may jeopardize the level of safety and therefore is at the user's own risk.

Kenneth Roy's *Safety First* column entries in *Science and Children* (in reverse chronological order):

- "Preventing Allergic Reactions," 53 (4): 27–29, December 2015.

- "Safety at First Sight," 53 (2): 93–95, October 2015.

- "Safer Science Explorations for Young Children," 53 (7): 26–28, March 2015.

- "Houston, We Have Liftoff!" 52 (3): 76–77, November 2014.

- "Feather, Feet, and Fin Safety in the Classroom," 52 (1): 72–74, September 2014.

- "Ensuring a Safer Outdoor Experience," 51 (7): 82–83, March 2014.

- "The Elementary Mission," 51 (2): 86–87, October 2013.

- "Getting Wired on Safety," 50 (7): 80–81, March 2013.

- "Modeling Safety in Clay Use," 50 (4): 84–85, December 2012.

Part I.

Teaching Science and Engineering in Early Childhood Settings

Chapter 1

Science Inquiry and the Use of *The Early Years* Column to Support Science and Engineering Learning

It can be daunting to know where to begin to teach science and engineering concepts at an age-appropriate level. Your students may be at many different levels of development and know much about one topic and little about another. I have been astonished at the amount of information one child had about polar bears, dismayed that another had never had the opportunity to hold a caterpillar, and gratified that both children were eager to make discoveries with the materials I made available for playing with light sources and creating shadows. No matter what experiences your children have had, science and engineering activities and the related discussions that you lead can help them learn more. By including science and engineering learning every day in your classroom, you are providing experiences that will prepare them for later learning and stimulate students' interest in learning mathematic and literacy skills as they seek to quantify and share their work. As a class endeavor, science and engineering learning will be the delight and responsibility of all your students, not only the few who have family support for such interest. As stated in *Taking Science to School,* a publication by The National Academies of Science Committee on Science Learning, Kindergarten Through Eighth Grade (NRC 2007, p. vii):

> *All young children have the intellectual capability to learn science. Even when they enter school, young children have rich knowledge of the natural world, demonstrate causal reasoning, and are able to discriminate between reliable and unreliable sources of knowledge. In other words, children come to school with the cognitive capacity to engage in serious ways with the enterprise of science.*

The Early Years is a monthly column in the National Science Teachers Association's (NSTA's) elementary school journal *Science and Children* that covers a wide range of topics in early childhood science. The column entries describe science and engineering activities designed to introduce children to, or to help children expand their understanding about, concepts that they often explore in their natural play. The activities engage children in the concepts described in the *National Science Education Standards* (NRC 1996) and the science and engineering practices described in *A Framework for K–12 Science Education: Practices, Crosscutting Concepts, and Core Ideas* (*Framework*; NRC 2012). The activities also align with the *Next Generation Science Standards* (*NGSS*; NGSS Lead States 2013), which draw from those two documents. In addition to recognizing the capabilities of young children, these documents call for educators to provide equitable opportunities for all students to engage in significant science and engineering learning (NRC 2012).

I am inspired by this quotation from the *Framework:*

> *The framework endeavors to move science education toward a more coherent vision in three ways. First, it is built on the notion of learning as a developmental progression. It is designed to help children continually build on and revise their knowledge and abilities, starting from their curiosity about what they see around them and their initial conceptions about how the world works. The goal is to guide their knowledge toward a more scientifically based and coherent view of the sciences and engineering, as well as of the ways in which they are pursued and their results can be used. (NRC 2012, pp. 10–11)*

The *Framework* also makes the following statement:

> *As in all inquiry-based approaches to science teaching, our expectation is that students will themselves engage in the practices and not merely learn about them secondhand. Students cannot comprehend scientific practices, nor fully appreciate the nature of scientific knowledge itself, without directly experiencing those practices for themselves. (NRC 2012, p. 2)*

Using Activities in the Process of Inquiry

As activities, the lessons in *The Early Years* column by themselves do not add up to science inquiry or an ongoing engineering design process. The column entries selected for this book are like recipes to use *when appropriate to support your students' experience of science inquiry and engineering design.* This book is a guide to activities that should be *part* of a larger science inquiry, just as a side dish is part of a meal that itself is part of a day's nutrition. When you select a particular activity, choose others that involve the same topic to develop the exploration into an inquiry (see Table 1.1, pp. 15–39). Science inquiry is an ongoing process through which children (and scientists) develop understanding. As described in the *National Science Education Standards*, inquiry

> *is a multifaceted activity that involves making observations; posing questions; examining books and other sources of information to see what is already known; planning investigations; reviewing what is already known in light of experimental evidence; using tools to gather, analyze, and interpret data; proposing answers, explanations, and predictions; and communicating the results. (NRC 1996, p. 23)*

"What are we doing in science this week?" is a question that suggests that teaching and learning science is a process designed to reveal facts to be added to a body of science content knowledge. Ask instead, "What *else* are we going to do with our exploration this week?" As children explore and investigate, guess and predict, and revisit ideas, they will learn facts *and* the ongoing process of science inquiry. Teachers in early childhood programs of all kinds should use the Early Years activities as a toolkit to find ways to develop a science inquiry about a concept or engineering design work or to build on the science inquiry and engineering design work that is taking place in their classroom. The goal is not to "cover" a large amount of material but to "uncover" children's ideas and build on them, an idea well developed in the Project Approach developed by Lilian Katz and Sylvia Chard (Katz and Chard 2000, p. 159) and the Reggio Emilia approach to education developed by Loris Malaguzzi and advocated for by Lella Gandini (Edwards, Gandini, and Forman 1998).

Because science is more than just doing many interesting things, you will do your students a disservice if you simply do these activities without connecting them as part of an ongoing discussion and exploration to find answers. "Science" is doing many interesting things to find out information and answer questions to build an understanding of the natural world. Finding out takes time, much thought about what happened, and doing more based on what you think may have happened. This book's collection of selected entries from *The Early Years* column, from September 2005 to Summer 2014, is most importantly a tool to use as you look for ways to help children focus on a single science concept or topic or engineering design problem over a lengthy period of time, even a few months in length.

Part III of this book includes all of the selected column entries in chronological order, as well as edited excerpts from and links to related *Early Years* blog posts. The column entries have been updated to include the *NGSS*, safety information, and additional resources. Table 1.1 lists all the column entries selected for this book, grouped by topics common to early childhood science and engineering curriculum.

Note that some column entries may be in more than one place in the table because they relate to more than one topic.

Use Table 1.1 to find activities that explore the same concepts and ideas in more than one way—a strategy to develop or extend science inquiry in your program. Select activities that align with your curriculum and state and national standards. Before you begin looking through the activities, remember that each activity is not a stand-alone science or engineering curriculum. Activities are small steps in a journey of science inquiry. Your students will understand concepts more deeply and learn more about the nature of science (NOS) if you consider how to be guided by their interests to build a series of activities into an ongoing exploration of a question, a concept, or a topic being investigated by your class. Ask yourself, "What should come before this and what should come after?"

The Nature of Science

Part of my teaching practice with young children is to talk about the NOS. I might say, "Just like scientists, you are trying to answer your question by making observations." Sometimes children ask difficult questions or make claims that are outside the realm of science, such as, "Are ghosts real?" We can refer to the "nature of science" and say, "As scientists we try to find out about the natural world, what we can observe. I have never seen, heard, felt, or smelled a ghost, except in fiction stories and movies about ghosts. Since I haven't observed ghosts, I can't study them in a scientific way."

Appendix H of the *NGSS* (NGSS Lead States 2013) presents the NOS basic understandings:

- Scientific investigations use a variety of methods.

- Scientific knowledge is based on empirical evidence.

- Scientific knowledge is open to revision in light of new evidence.

- Scientific models, laws, mechanisms, and theories explain natural phenomena.

- Science is a way of knowing.

- Scientific knowledge assumes an order and consistency in natural systems.

- Science is a human endeavor.

- Science addresses questions about the natural and material world.

These understandings are presented in a matrix with grade-level understandings about the NOS. This matrix is very useful in understanding the learning outcomes about the NOS that can be expected by grade 2. Children begin forming ideas about the natural world well before kindergarten. We can support their continued refinement of these ideas through experience with science inquiry based on the nature of science. This way of thinking is important to many areas of learning throughout life (Worth and Grollman 2003).

To learn more about the practices and nature of science, see the University of California Museum of Paleontology's website Understanding Science: How Science Really Works (*http://undsci.berkeley.edu*), especially the "Understanding Science 101" section.

Engineering in Early Childhood

There is overlap in the focus on engineering concepts and science concepts in some activities in *The Early Years* column. However, the main emphasis of the column and related blog posts is on activities that can be part of a science inquiry. Refer to the *Framework* (NRC 2012), the *NGSS* (NGSS Lead States 2013), and engineering resources for more information on how to expand the use of engineering in early childhood (see Appendix 1, pp. 345–346, for a list of additional resources for science inquiry and engineering design).

How Is Science Inquiry Different From Doing a Science Activity? An Example at Two Age Ranges

A science activity engages children in a hands-on session that may happen once or a few times. Children generally enjoy exploring new materials. But, as educator and researcher Jeff Winokur puts it, "Just because they see it [a phenomenon] doesn't mean they understand it" (Ashbrook 2012). Documenting their observations, taking part in discussions to reflect on their work, and follow-up work after the initial hands-on experiences can deepen children's understanding of the concepts.

NSTA's position statement on early childhood science education (see Appendix 2) recognizes that "adults play a central and important role in helping young children learn science" (see also Worth 2014). By observing your students to find out what interests them, and making changes in materials to extend their exploration, you intentionally plan a learning environment where ongoing inquiry develops. By asking open-ended questions, listening to students' ideas, and discussing their reasoning, you encourage children to expand their understanding.

In this section I give an example of how an activity, playing with wet sand, can be part of a broader inquiry involving science and engineering concepts about the properties of water. The activity is described for two age ranges of typically developing children: ages 3–6 years and ages 5–8 years (see Boxes 1.1 and 1.2, p. 8). Note that there is overlap in the age ranges. This is because young children's development can widely vary from child to child, and between abilities within one child. Children's prior experiences also determine where they begin and develop an exploration.

Within the daily routine of a prekindergarten (preK)–grade 2 school day, over the course of months children have already been engaged in exploring and thinking about the properties of water by activities such as wetting bandannas and hanging them to dry, using small pumps to pull water up out of buckets and pouring water back in, using misters to keep a terrarium damp, and freezing water and melting ice. They have had many hours of open-ended building with and digging in sand in an outdoor sandbox and an inside sensory table. The children have also been drawing and writing about what they see and think, and talking about their ideas with each other and with a teacher. Now they will experience how one property of a liquid, surface tension, acts to hold sand together.

BOX 1.1. Wet Sand Activity for Children Ages 3–6 Years

This set of possible conversations and actions probably won't happen in a single sequence. Here it is written without all the interruptions (for trips to the bathroom, helping to resolve conflicts, and stopping for the end of the day). The words and actions will be different each time it happens, even with the same teacher.

1. *You notice several children at the sensory table piling up dry sand.*

2. *You ask an open-ended question, "Tell me what you are working on here."*

3. *The children tell you that they are trying to build a sand castle but "it's not working."*

---

4. You ask, "What else have you tried?"

5. The children say they have filled cups and turned them over but the sand "doesn't stay up."

6. You say, "Let's draw what you can do with this sand," and provide drawing materials (perhaps one group drawing at a table, or individual drawings on small clipboards). The children's representations may not look like loose piles of sand, but by writing down their dictation you can capture more of their thinking. They may be able to write their names and the word "sand" on this documentation of observation.

7. You suggest, "Let's look at the sand using magnifiers to see what is happening to the sand grains when you pile them up." The children look closely and see that the sand grains look like little rocks, that they roll, and that there are "too many to count."

8. You ask, "I wonder what could make sand stay up?" Children with prior experience may be able to remember that sand needs to be wet to stick together for building and may ask for water in the sensory table. Children without that experience will probably need additional experience, a direct suggestion from a teacher or another child, or to learn about others' experiences through discussion or listening to a book (that mentions and pictures building sand castles at the beach).

9. You ask the children to think of and discuss possible actions. Write down the children's thoughts because this is an important part of the inquiry and design process—talking about what might work to solve a problem or discover an answer and reflecting back on it after trying it. This might only take a few minutes.

10. You help the children act on their ideas, which may include using water. At a table, give them materials that will allow them to focus on trying their various ideas on a small scale: about a tablespoon of sand in each bowl; and other materials such as cups of water and droppers, liquid school glue, sticky rice, spoons, tape, pipe cleaners, and magnifiers. The children make a mess but also make meaning out of their exploration of the provided materials. Most children will not like the glue and sand sticking to their fingers, so although that might be a workable solution, they will not want glue in the sensory table. They will enjoy trying what happens with the other materials but will settle on adding water to the sand to make it stick

together. Some will get distracted and begin building with the tape and pipe cleaners, discovering properties of those materials that will help them solve different problems in the future. These side explorations are not "off task" but another way to learn about using materials.

11. You ask, "How do you think we can find out how much water to add to the sensory table?" and start the children on another part of the inquiry and design process, mixing different proportions of sand to water and comparing the results. Some children will prefer an open-ended exploration of adding water and sand "as needed" in a sensory table or sandbox. How carefully they measure and combine will depend on their individual development. You can provide limited sand and water, cups, small measuring tools (small spoons and shot glasses, pill cups, or graduated 1-oz measures), and help the children record both their findings and their thoughts on why one amount of sand and water can build a taller mound than other amounts. Discussion as the water and sand are combined will help children reflect on what they observe and experience. Refer back to the documentation of their work with dry sand and ask children to explain their understanding of how wet sand is a better building material.

BOX 1.2. Wet Sand Activity for Children Ages 5–8 Years

This set of possible conversations and actions is similar but not identical to the set for children ages 3–6 years, and these conversations and actions also are unlikely to happen in a single sequence. Here it is written without all the interruptions (for trips to the bathroom, helping others with spelling, breaking for lunch, and stopping for the end of the day). The words and actions will be different each time it happens, even with the same teacher.

1. You notice several children at the sensory table piling up dry sand.

2. You ask an open-ended question, "Tell me what you are working on here."

3. The children tell you that they are trying to build a sand castle but "it's not working, and we need to add water." These slightly older children are more likely to have experienced the difference in building with wet and dry sand.

4. You ask, "What else have you tried?"

5. The children say they have filled cups and turned them over but the sand "doesn't stay up, so we need water."

6. You say, "Let's look at the chart to see what we already know about water." (The class is engaged in a science inquiry about the properties of water.) You and the children read from a graphic organizer, such as a simple list of observations or a K-W-L chart (what I think I KNOW—what I WANT to know—what I LEARNED): Water makes things wet. Water is wet. It is a liquid. But it can be a solid if it's ice. Water evaporates. Plants need water. I drink water. I use water to wash my hands. I use a paper towel to dry my hands. Water stays on my hands after I wash them. Water pours. Rain is water. Snow is water.

7. You suggest, "Before we add water, let's look at the sand using magnifiers to see what is happening to the sand grains when you pile them up." The children look closely and draw, dictate to you, and write to document their observations, "The sand doesn't stay together."

8. These children are ready for a quantitative activity to discover how much water to add to sand to make a mixture that is good for building, because their prior experience suggests that sand needs to be wet to stick together for building.

9. You ask, "How much water do you want to add?" and, if needed after some discussion, ask, "What could you do to find out the best amount of water to add?" Write down the children's thoughts because this is an important part of inquiry—talking about what might work to solve a problem or discover an answer.

10. This starts the children on another part of the inquiry, mixing different proportions of sand to water and comparing the results. How carefully they measure and combine will depend on their individual development.

11. You provide limited sand and water, cups, and small measuring tools (small spoons and shot glasses, pill cups, or graduated 1-oz measures) and help the children keep track of how much of each material they mix. Children can record their findings through drawing or graphing, and they can write or dictate their thoughts on why one amount of sand and water can build a taller mound than other amounts. When there are disagreements, tell the children that scientists and engineers may have different ideas. Ask the children to state their ideas and the evidence for why they think that. What have they noticed or experienced with wet sand that supports their idea?

12. The children may agree on an amount of water, but you should have additional dry sand available in case they find they have overwatered the sand.

These example activities are part of a broader inquiry into the properties of water. These hands-on activities will help the children understand in later years how the molecular structure of water makes building with wet sand possible. An ongoing inquiry about the properties of water can include experiences with freezing and melting, evaporation, dissolving solids in water, and creating flow in water systems. This inquiry should take place over months, with adult guidance providing materials, time for reflection, and discussion.

Planning Science Inquiry

When planning to teach science, ask yourself what you want the children to learn. There are many fun activities to try, but not all of them will be part of an ongoing investigation that engages children in thinking about the same core idea or concept over a length of time. Being involved in exploring a concept over time allows children to consider their thinking and revise ideas that do not agree with the evidence of their investigation. Observe your students to see what interests them and what questions they have, and review your state and national standards to determine age-appropriate concepts that align with your students' desire to know more. The *National Science Education Standards* for grades K–4 (NRC 1996), *A Framework for K–12 Science Education* (NRC 2012), the Project 2061 *Benchmarks for Science Literacy* (AAAS 1993), and the *NGSS* (NGSS Lead States 2013) offer guidance in choosing age-appropriate concepts to address. Developed with much consideration and debate by national experts, these standards address science learning in grades K–12. Preschool educators in all programs can use them to see what conceptual learning their children will be ready for in kindergarten. Given the range of development in early childhood, some preschool children are ready to understand the concepts in kindergarten standards, but there is no reason to try to have all students reach early competency. Children's experiences before kindergarten are an essential foundation for later understanding of the crosscutting concepts and disciplinary core ideas described in the *NGSS* and the performance expectations for elementary school and beyond. State standards for preK can be used together with these national standards to choose topics and concepts to investigate with your class.

NSTA's position statement on early childhood science education (see Appendix 2, pp. 347–352) is a document specifically for teachers of preschool children. Regardless of age, children need time to explore the materials, discuss their thinking, and try new arrangements of the materials (among other actions), to be able to learn as much as they are developmentally ready to learn about a concept or topic (Chalufour and Worth 2006). For additional guidance in planning early childhood learning (up to age 8), see position statements and other guidance by professional associations such as the National Association for the Education of Young Children (NAEYC; see *www.naeyc.org/positionstatements*) and the Association for Constructivist Teaching (see *https://sites.google.com/site/assocforconstructteaching*) and program standards such as the Head Start and Early Learning Outcomes Framework (U.S. DHHS, ACF, Office of Head Start 2015). Think about how you will assure that all students will be able to use the materials and understand your directions and questions. Refer to the position statement on early childhood inclusion (Division for Early Childhood of the Council for Exceptional Children / NAEYC 2009) for guidance. Look for resources that will support students who are dual language learners. Science explorations that connect to children's life experiences help all children build on their prior knowledge.

Plan to connect the science inquiry with other areas of the curriculum, just as science in the greater world connects to our daily life, by incorporating the use of mathematics and early literacy skills as a natural part of the work. Children want to answer how many and how much in their science inquiry. Once they recognize that print has meaning, they want to use it in their own work to share their thoughts. Science inquiry motivates children to find out more by measuring and collecting data and by reading and looking at pictures. They want to explain discoveries themselves through mathematics and writing.

It is easy to be passionate about teaching science and engineering in early childhood classrooms when one considers how far-reaching the experiences will be for the children. Promoting science learning by facilitating science inquiry in grades preK and up enriches children's lives by broadening their base of experiences and providing opportunities to think and talk about what they see happening, learn new vocabulary and new skills, develop an understanding of the reasons for measurement, and use their beginning writing and drawing skills to communicate their thinking. Providing opportunities for children to design solutions to problems and test them empowers children to test new ideas and make changes to their solutions. Science inquiry engages children in work where questions are valued because asking questions is central to the work. As children work together in activities to explore a question, they build social skills. Science inquiry and engineering design promote working with others, and children's literature can be used to add to what children know already. As activities build toward an ongoing inquiry, children can reflect on their previous ideas, sometimes correcting a misconception.

Few of the activities in this book are new. Who was the first teacher to use plastic sandwich bags to hold sprouting seeds, or the first person to make a kazoo out of a cardboard tube and wax paper? The step-by-step organization and the listed resources will help you implement these activities, which have been used in my preschool classrooms with children ages 2 to 5 years old since 1998. It is important to note that these activities should not be used alone or in a random order because this may lead to a disjointed view or incomplete understanding of science and engineering concepts. Instead, these activities should be used in the service of a sustained science inquiry or engineering design process.

Science activities become part of a strong science curriculum when they are implemented as part of a thoughtfully planned science inquiry about an investigable question. An in-depth science inquiry is an ongoing investigation where children are introduced to materials through hands-on experiences and, with teacher guidance, begin to investigate a question that they can answer through their own actions, observations, and with teacher-assisted research. Qualities that make an experience appropriate to include in early childhood science inquiry are described as being interesting to children, linked to their experiences, and accessible to all children's direct exploration (Chalufour and Worth 2006; Harlen and Qualter 2009).

Not all activities I propose are embraced by my students—what is successful in sustaining an investigation at one age, or in one classroom, provokes little interest in another. Your careful observation of your students at work and the standards followed by your program will guide your choices. This book will allow you to pick and combine activities to build science inquiry into a topic or a concept. Each year you teach you will build your science "kits" of personal experience and materials, including books related to favorite topics, adding a few each year as your budget allows. Each year you will be better prepared to facilitate the science inquiry your students are capable of engaging in.

Being a Beginner in the Process of Inquiry

Can young children analyze data? Are you prepared to guide them in their investigations to experience the world and gather data? If you don't feel ready, don't be afraid or embarrassed. Being a beginner is fine if you are open to learning how to do more or to do it more accurately. We all have been beginners. From my work as a family home-childcare provider, I knew that young children are capable of noticing tiny details, sharing their interest in small animals, and understanding fairly complex models. My teaching moved toward child-directed exploration as I saw how the connections they made in their play supported what we talked about in conversation, and I relaxed, no longer feeling that I had to teach them everything. Learning from others (see resources listed in Appendix 1, pp. 345–346) broadened my understanding of science teaching to include establishing science inquiry in the classroom, expanding observation activities, and documenting children's work and asking what meaning they make from it. You can do this, too, if you are not already doing it, and you can use this book to find additional ideas for implementing science activities in the course of science inquiry in your classroom.

Take note when children show interest by their actions or words in natural phenomena such as the flow of water or the lives of insects. Extend their investigations by providing additional time and materials to increase their observations. Use conversations to find out what the children currently understand so you can support their continuing development of concept understanding (Sander and Nelson 2009). Engage the children with an activity about a concept such as "water will flow down toward the ground" or "animals have specific needs to grow and reproduce that are met by particular environments." The initial activity could be as simple as playing with water in a water table or observing a beetle. Observe the students. What aspect(s) are they interested in? Engage them in conversation, writing and drawing exercises, and group discussion to find out what they observed and what questions they have. Focus the observations and analysis of data through group and individual conversations, recording children's statements to refer to in further discussions. Use reflection on previous experiences to help children share their evidence for their ideas and possibly revise their thinking. Do allow children to work without direct instruction for a time, struggling to solve their own difficulties by creating their own solutions. Resist the temptation to share all you know about the topic they are investigating, whether it is building a sturdy tower or looking for the best place to find worms. Instead, ask yourself, "What additional materials can I introduce in the classroom to expand the students' work?"

This is all easier said than done, but it becomes easier with practice. Reach out to colleagues in your program or professional association, such as the NAEYC Early Childhood Science Interest Forum (*http://member-forums.naeyc.org/group/early-childhood-science-interest-forum*), for support and resources. We can allow ourselves time to build our ability to teach science through practice, reading resources, and taking advantage of professional development.

References

American Association for the Advancement of Science (AAAS). 1993. *Benchmarks for science literacy*. New York: Oxford University Press.

Ashbrook, P. 2012. Early learning experiences build toward understanding concepts that are hard to teach. Early Years [blog]. *http://nstacommunities.org/blog/2012/09/27/early-learning-experiences-build-toward-understanding-concepts-that-are-hard-to-teach*.

Chalufour, I., and K. Worth. 2006. Science in kindergarten. In *K today: Teaching and learning in the kindergarten year*, ed. D. F. Gullo, 95–106. Washington, DC: National Association for the Education of Young Children (NAEYC). Also available online as Reading #56: Science in kindergarten. A reading from the CD accompanying *Developmentally appropriate practice in early childhood programs serving children from birth through age 8*, 3rd ed., eds. C. Copple and S. Bredekamp. Washington, DC: NAEYC, 2009. *www.rbaeyc.org/resources/Science_Article.pdf*.

Copple, C., and S. Bredekamp, eds. 2009. *Developmentally appropriate practice in early childhood programs serving children from birth through age 8.* 3rd ed. Washington, DC: National Association for the Education of Young Children.

Division for Early Childhood of the Council for Exceptional Children/National Association for the Education of Young Children (NAEYC). 2009. *Early childhood inclusion: A joint position statement of the Division for Early Childhood (DEC) and the National Association for the Education of Young Children (NAEYC).* Chapel Hill: The University of North Carolina, FPG Child Development Institute. *www.naeyc.org/files/naeyc/file/positions/DEC_NAEYC_EC_updatedKS.pdf.*

Edwards, C., L. Gandini, and G. Forman, eds. 1998. *The hundred languages of children: The Reggio Emilia approach—Advanced reflections.* 2nd ed. Westport, CT: Ablex.

Harlen, W., and A. Qualter. 2009. *The teaching of science in primary schools.* 5th ed. London: David Fulton.

Katz, L., and S. Chard. 2000. *Engaging children's minds: The Project Approach.* Stamford, CT: Ablex.

National Research Council (NRC). 1996. *National science education standards.* Washington, DC: National Academies Press.

National Research Council (NRC). 2007. *Taking science to school: Learning and teaching science in grades K–8,* Committee on Science Learning, Kindergarten Through Eighth Grade; eds. R. A. Duschl, H. A. Schweingruber, and A. W. Shouse. Washington, DC: National Academies Press.

National Research Council (NRC). 2012. *A framework for K–12 science education: Practices, crosscutting concepts, and core ideas.* Washington, DC: National Academies Press. *www.nap.edu/catalog/13165/a-framework-for-k-12-science-education-practices-crosscutting-concepts.*

NGSS Lead States. 2013. *Next Generation Science Standards: For states, by states.* Washington, DC: National Academies Press. *www.nextgenscience.org/next-generation-science-standards.*

Sander, J., and S. Nelson. 2009. Science conversations for young learners. *Science and Children* 46 (6): 43–45.

U.S. Department of Health and Human Services (DHHS), Administration for Children and Families (ACF), Office of Head Start. 2015. *Head Start early learning outcomes framework: Ages birth to five.* Washington, DC: DHHS, ACF, Office of Head Start. *http://eclkc.ohs.acf.hhs.gov/hslc/hs/sr/approach/pdf/ohs-framework.pdf.*

Worth, K. 2014. Science in early childhood education. Interviewed by Brian Bartel. *Lab Out Loud,* Episode 108, February 24. *http://laboutloud.com/2014/02/episode-108-science-in-early-childhood-education.*

Worth, K., and S. Grollman. 2003. *Worms, shadows, and whirlpools: Science in the early childhood classroom.* Portsmouth, NH: Heinemann.

TABLE 1.1. The Early Years selected column entries by topic, area of science, and connections to the *Framework* and the *NGSS*

Chapter number; original publication date	Title of column entry/Activity	Area of Science					*Framework* and *NGSS* connections
		Physical sciences*	Chemical sciences†	Life sciences	Earth and space sciences	NOS and nature of scientific inquiry	
Nature of Science and Science Process Skills							
14; September 2005	What Can Young Children Do as Scientists?/Scientists at Work					•	*Framework* All practices *NGSS* Appendix H (Science is a human endeavor)
18; September 2006	Young Questioners/Name That Object					•	*Framework* Practice: Asking questions *NGSS* Appendix H (Scientific knowledge is based on empirical evidence)
19; October 2006	Learning Measurement/Measuring Hands					•	*Framework* Practice: Using mathematics and computational thinking *NGSS* Appendix H (Scientific investigations use a variety of methods)

Table 1.1 (continued)

Nature of Science and Science Process Skills

Chapter number; original publication date	Title of column entry/Activity	Area of Science					Framework and NGSS connections
		Physical sciences*	Chemical sciences†	Life sciences	Earth and space sciences	NOS and nature of scientific inquiry	
26; February 2008	Observing With Magnifiers/Exploring Magnifiers	•				•	*Framework* DCI: PS4.B *NGSS* PEs: K-2-ETS1-2, 1-PS4-3 Appendix H (Scientific investigations use a variety of methods; scientific knowledge is based on empirical evidence)
37; March 2010	Building With Sand/How High Can You Build With Sand?				•	•	*Framework* Practice: Using mathematics and computational thinking; constructing explanations and designing solutions DCIs: PS1.A, ETS1.C *NGSS* PEs: 2-PS1-2, K-2-ETS1-3 Appendix H (Scientific investigations use a variety of methods; scientific knowledge is based on empirical evidence)
38; September 2010	Inquiry at Play/A Scientist Visits					•	*NGSS* Appendix H (all)

Table 1.1 (continued)

Chapter number; original publication date	Title of column entry/Activity	Area of Science					Framework and NGSS connections
		Physical sciences*	Chemical sciences†	Life sciences	Earth and space sciences	NOS and nature of scientific inquiry	
Nature of Science and Science Process Skills							
39; October 2010	Developing bservation Skills/Making and Observing Bubbles	•				•	*Framework* Practices: Asking questions and defining problems; planning and carrying out investigations *NGSS* PE: 2-PS1-1 Appendix H (Scientific knowledge is based on empirical evidence)
40; December 2010	Investigable Questions/How Much Is Enough?			•		•	*Framework* Practices: Asking questions; planning and carrying out investigations; analyzing and interpreting data *NGSS* PEs: K-LS1, 2-LS2 Appendix H (Scientific investigations use a variety of methods)

Table 1.1 (*continued*)

Chapter number; original publication date	Title of column entry/Activity	Area of Science					Framework and NGSS connections
		Physical sciences*	Chemical sciences†	Life sciences	Earth and space sciences	NOS and nature of scientific inquiry	
Nature of Science and Science Process Skills							
42; April 2011	Sharing Research Results/Playdough Guidebook		•			•	*Framework* Practices: Analyzing and interpreting data; obtaining, evaluating, and communicating information DCIs: PS1.A, PS1.B *NGSS* PEs: 2-PS1-1, 2-PS1-2, 2-PS1-4 Appendix H (Scientific knowledge is based on empirical evidence; scientific knowledge is open to revision in light of new evidence)
44; September 2011	A Sense of Place/Schoolyard as Model				•		*Framework* Practice: Developing and using models CCCs: Scale, proportion, and quantity; systems and system models *NGSS* PE: 2-ESS2-2 Appendix H (Scientific models, laws, mechanisms, and theories explain natural phenomena)

Table 1.1 (continued)

Chapter number; original publication date	Title of column entry/Activity	Area of Science					Framework and NGSS connections
		Physical sciences*	Chemical sciences†	Life sciences	Earth and space sciences	NOS and nature of scientific inquiry	
Nature of Science and Science Process Skills							
45; November 2011	Reading Stories and Making Predictions/"What Might Happen If ...?"					•	*Framework* Practices: Asking questions; constructing explanations and designing solutions; engaging in argument from evidence CCC: Cause and effect: Mechanism and explanation *NGSS* Appendix H (Scientific investigations use a variety of methods)
46; January 2012	Seeing the Moon/ Crater Making	•			•	•	*Framework* Practice: Developing and using models CCC: Cause and effect: Mechanism and explanation *NGSS* PE: 1-ESS1-1 Appendix H (Scientific investigations use a variety of methods; scientific knowledge is based on empirical evidence; scientific knowledge assumes an order and consistency in natural systems)

Table 1.1 (continued)

Chapter number; original publication date	Title of column entry/Activity	Area of Science					Framework and NGSS connections
		Physical sciences*	Chemical sciences†	Life sciences	Earth and space sciences	NOS and nature of scientific inquiry	
Nature of Science and Science Process Skills							
47; July 2012	Circle Time/Teach and Tell					•	*Framework* Practices: Engaging in argument from evidence; obtaining, evaluating, and communicating information *NGSS* Appendix H (Science is a human endeavor)
49; December 2012	Please Touch Museum/Materials Museum	•				•	*Framework* DCI: PS1.A *NGSS* PE: 2-PS1-1 Appendix H (Scientific knowledge assumes an order and consistency in natural systems; science is a human endeavor)
57; September 2014	The Nature of Science in Early Childhood/Observations and Inferences					•	*Framework* Practice: Engaging in argument from evidence *NGSS* Appendix H (all)

Table 1.1 (continued)

Objects in Motion

Chapter number; original publication date	Title of column entry/Activity	Area of Science					Framework and NGSS connections
		Physical sciences*	Chemical sciences†	Life sciences	Earth and space sciences	NOS and nature of scientific inquiry	
17; July 2006	Roll With It/Wheel Work	•					*Framework* Practices: Planning and carrying out investigations; constructing explanations (for science) and designing solutions (for engineering). CCC: Structure and function. *NGSS* PEs: K-PS2, K-2-ETS1-2
24; July 2007	Water Works/How Can We Move Water?	•	•		•		*Framework* DCIs: PS2.A, ETS1.A, ETS1.B *NGSS* PEs: K-PS2, K-2-ETS1-3
27; March 2008	Objects in Motion/Crash Dummy Fun!	•					*Framework* DCI: PS2.A *NGSS* PEs: K-PS2-1, K-PS2-2, K-2-ETS1-3
29; December 2008	Air Is Not Nothing/Moving Air		•				*Framework* DCI: PS1.A *NGSS* PEs: K-PS2-1, 1-PS4-1
32; April 2009	Hear That?/Make a Kazoo	•					*Framework* DCI: PS4.C *NGSS* PEs: K-PS2, 1-PS4-1

1

Table 1.1 (continued)

Chapter number; original publication date	Title of column entry/Activity	Area of Science					Framework and NGSS connections
		Physical sciences*	Chemical sciences†	Life sciences	Earth and space sciences	NOS and nature of scientific inquiry	
Objects in Motion							
41; February 2011	Ongoing Inquiry/ Properties of Water	•					*Framework* All practices
43; July 2011	Measuring Learning/ Temperature Changes	•					*Framework* Practice: Using mathematics and computational thinking *NGSS* PE: K-ESS2-1
46; January 2012	Seeing the Moon/ Crater Making	•			•	•	*Framework* Practice: Developing and using models *NGSS* PE: 1-ESS1-1
48; November 2012	Drawing Movement/ Just Draw It—Ramps	•					*Framework* Practices: Planning and carrying out investigations; designing solutions *NGSS* PEs: K-PS2-1, K-PS2-2
What Are Things Made Of?							
15; January 2006	The Matter of Melting/ Melt Away!	•	•				*Framework* DCI: PS1.A *NGSS* PEs: 2-PS1-1, 2-PS1-4, 2-ESS2-3

Table 1.1 (continued)

Chapter number; original publication date	Title of column entry/Activity	Area of Science					Framework and NGSS connections
		Physical sciences*	Chemical sciences†	Life sciences	Earth and space sciences	NOS and nature of scientific inquiry	
What Are Things Made Of?							
20; December 2006	Rocks Tell a Story/Pretend Rocks				•		*Framework* DCI: PS1.A *NGSS* PEs: 2-PS1-1, 2-PS1-3, 2-ESS1-1
26; February 2008	Observing With Magnifiers/Exploring Magnifiers	•				•	*Framework* DCI: PS4.B *NGSS* PE: K-2-ETS1-2
29; December 2008	Air Is Not Nothing/Moving Air		•				*Framework* DCI: PS1.A *NGSS* PE: K-PS2-1, 1-PS4-1
31; March 2009	Does Light Go Through It?/Playing With Light	•					*Framework* DCI: PS4.B *NGSS* PEs: 1-PS4-3, 2-PS1-1
37; March 2010	Building With Sand/How High Can You Build With Sand?				•	•	*Framework* Practice: Constructing explanations and designing solutions DCIs: PS1.A, ETS1.C *NGSS* PEs: 2-PS1-2, K-2-ETS1-3

Table 1.1 (continued)

Chapter number; original publication date	Title of column entry/Activity	Area of Science					Framework and NGSS connections
		Physical sciences*	Chemical sciences†	Life sciences	Earth and space sciences	NOS and nature of scientific inquiry	
What Are Things Made Of?							
39; October 2010	Developing Observation Skills/Making and Observing Bubbles	•				•	*Framework* Practices: Asking questions and defining problems; planning and carrying out investigations *NGSS* PE: 2-PS1-1
41; February 2011	Ongoing Inquiry/Properties of Water	•					*Framework* All practices
42; April 2011	Sharing Research Results/Playdough Guidebook	•	•			•	*Framework* DCIs: PS1.A, PS1.B *NGSS* PEs: 2-PS1-1, 2-PS1-2, 2-PS1-4
49; December 2012	Please Touch Museum/Materials Museum	•				•	*Framework* DCI: PS1.A *NGSS* PE: 2-PS1-1
54; November 2013	Are They Getting It?/Materials Have Properties	•				•	*Framework* DCIs: PS1.A, ETS1 *NGSS* PEs: 2-PS1, K-2-ETS1
56; July 2014	Inviting Parents to School/Making Lemonade From Scratch	•	•				*Framework* DCI: PS1.A *NGSS* PEs: 2-PS1-1, 2-PS1-2

Table 1.1 (continued)

Chapter number; original publication date	Title of column entry/Activity	Area of Science					Framework and NGSS connections
		Physical sciences*	Chemical sciences†	Life sciences	Earth and space sciences	NOS and nature of scientific inquiry	
Mixing and Making a Change							
15; January 2006	The Matter of Melting/Melt Away!	•	•				*Framework* DCI: PS1.A *NGSS* PEs: 2-PS1-1, 2-PS1-4, 2-ESS2-3
20; December 2006	Rocks Tell a Story/Pretend Rocks				•		*Framework* DCI: PS1.A *NGSS* PEs: 2-PS1-1, 2-PS1-3, 2-ESS1-1
29; December 2008	Air Is Not Nothing/Moving Air		•				*Framework* DCI: PS1.A *NGSS* PE: K-PS2-1, 1-PS4-1
37; March 2010	Building With Sand/How High Can You Build With Sand?		.		•	•	*Framework* Practice: Constructing explanations and designing solutions DCIs: PS1.A, ETS1.C *NGSS* PEs: 2-PS1-2, K-2-ETS1-3
42; April 2011	Sharing Research Results/Playdough Guidebook		•			•	*Framework* DCIs: PS1.A, PS1.B *NGSS* PEs: 2-PS1-1, 2-PS1-2, 2-PS1-4

Table 1.1 (continued)

Chapter number; original publication date	Title of column entry/Activity	Area of Science					Framework and NGSS connections
		Physical sciences*	Chemical sciences†	Life sciences	Earth and space sciences	NOS and nature of scientific inquiry	
Mixing and Making a Change							
55; December 2013	Now We're Cooking/ Ice Cream Science	•					*Framework* DCIs: PS1.A, PS1.B *NGSS* PE: 2-PS-1
56; July 2014	Inviting Parents to School/Making Lemonade From Scratch	•	•				*Framework* DCI: PS1.A *NGSS* PEs: 2-PS1-1, 2PS1-2
Exploring Our Senses							
16; April 2006	Feet First/Footprint Fun			•			*Framework* CCC: Structure and function DCI: LS1.A *NGSS* PEs: K-2-ETS1-2, 1-LS1-1
22; March 2007	The Sun's Energy/ Making Sun Prints				•		*Framework* DCI: PS3.B *NGSS* PE: K-PS3-1
26; February 2008	Observing With Magnifiers/Exploring Magnifiers	•				•	*Framework* DCI: PS4.B *NGSS* PE: K-2-ETS1-2

Table 1.1 (continued)

Chapter number; original publication date	Title of column entry/Activity	Area of Science					Framework and NGSS connections
		Physical sciences*	Chemical sciences†	Life sciences	Earth and space sciences	NOS and nature of scientific inquiry	
Exploring Our Senses							
31; March 2009	Does Light Go Through It?/Playing With Light	•					*Framework* DCI: PS4.B *NGSS* PEs: 1-PS4-3, 2-PS1-1
32; April 2009	Hear That?/Make a Kazoo	•					*Framework* DCI: PS4.C *NGSS* PEs: K-PS2, 1-PS4-1
35; October 2009	Safe Smelling/Do You Smell What I Smell?		•	•			*Framework* DCIs: LS1.D, PS4.C *NGSS* PE: 1-LS1-1
42; April 2011	Sharing Research Results/Playdough Guidebook	•				•	*Framework* DCIs: PS1.A, PS1.B *NGSS* PEs: 2-PS1-1, 2-PS1-2, 2-PS1-4
43; July 2011	Measuring Learning/Temperature Changes	•					*Framework* Practice: Using mathematics and computational thinking *NGSS* PE: K-ESS2-1
49; December 2012	Please Touch Museum/Materials Museum	•				•	*Framework* DCI: PS1.A *NGSS* PE: 2-PS1-1

Table 1.1 (continued)

Chapter number; original publication date	Title of column entry/Activity	Area of Science					Framework and NGSS connections
		Physical sciences*	Chemical sciences†	Life sciences	Earth and space sciences	NOS and nature of scientific inquiry	
Exploring Our Senses							
56; July 2014	Inviting Parents to School/Making Lemonade From Scratch	●	●				*Framework* DCI: PS1.A *NGSS* PEs: 2-PS1-1, 2-PS1-2
Learning About Plants							
22; March 2007	The Sun's Energy/Making Sun Prints				●		*Framework* DCI: PS3.B *NGSS* PE: K-PS3-1
23; April 2007	Collards and Caterpillars/What Do Caterpillars Eat?			●			*Framework* DCIs: LS1.A, LS1.B, LS1.C, LS2.A, LS2.B, LS2.C *NGSS* PEs: K-LS1-1, 1-LS1-2
28; July 2008	An Invertebrate Garden/Attracting Invertebrates			●			*Framework* DCIs: LS1.A, LS1.B, LS1.C, LS2.A, LS2.B, LS2.C *NGSS* PEs: K-LS1-1, 1-LS1-2
30; February 2009	Bring On Spring: Planting Peas/Planting Peas and Observing Growth			●			*Framework* DCI: LS1.A *NGSS* PEs: K-LS1-1, 1-LS1-2, 2-LS2-1

Table 1.1 (continued)

Chapter number; original publication date	Title of column entry/Activity	Area of Science					Framework and NGSS connections
		Physical sciences*	Chemical sciences†	Life sciences	Earth and space sciences	NOS and nature of scientific inquiry	
Learning About Plants							
34; September 2009	Planting Before Winter/Some Like It Cold			•			*Framework* DCI: LS1.C *NGSS* PE: K-LS1
35; October 2009	Safe Smelling/Do You Smell What I Smell?			•			*Framework* DCIs: LS1.D, PS4.C *NGSS* PE: 1-LS1-1
40; December 2010	Investigable Questions/How Much Is Enough?			•		•	*Framework* Practices: Asking questions; planning and carrying out investigations; analyzing and interpreting data *NGSS* PEs: K-LS1-1, 2-LS2-1
49; December 2012	Please Touch Museum/Materials Museum	•				•	*Framework* DCI: PS1.A *NGSS* PE: 2-PS1-1
51; February 2013	"Life" Science/A Plant's Life			•			*Framework* DCIs: LS1.A, LS1.B, LS1.C *NGSS* PE: 3-LS1

Table 1.1 (continued)

Chapter number; original publication date	Title of column entry/Activity	Area of Science					Framework and NGSS connections
		Physical sciences*	Chemical sciences†	Life sciences	Earth and space sciences	NOS and nature of scientific inquiry	
Learning About Animals							
16; April 2006	Feet First/Footprint Fun			•			*Framework* CCC: Structure and function DCI: LS1.A *NGSS* PEs: K-2-ETS1-2, 1-LS1-1
21; February 2007	Birds in Winter/Bird Shapes			•			*Framework* DCIs: LS1-A, LS1-C, LS2-A *NGSS* PEs: K-LS1-1, KESS3-1, 2-LS4-1
23; April 2007	Collards and Caterpillars/What Do Caterpillars Eat?			•			*Framework* DCIs: LS1.A, LS1.B, LS1.C, LS2.A, LS2.B, LS2.C *NGSS* PEs: K-LS1-1, 1-LS1-2
25; October 2007	Counting a Culture of Mealworms/Beetle Roundup			•			*Framework* Practices: Using mathematics and computational thinking; analyzing and interpreting data; engaging in argument from evidence DCIs: LS1.A, LS1.C *NGSS* PEs: K-LS1-1, 1-LS3-1

Table 1.1 (continued)

Chapter number; original publication date	Title of column entry/Activity	Area of Science					Framework and NGSS connections
		Physical sciences*	Chemical sciences†	Life sciences	Earth and space sciences	NOS and nature of scientific inquiry	
Learning About Animals							
28; July 2008	An Invertebrate Garden/Attracting Invertebrates			•			*Framework* DCIs: LS1.A, LS1.B, LS1.C, LS2.A, LS2.B, LS2.C *NGSS* PEs: K-LS1-1, 1-LS1-2
35; October 2009	Safe Smelling/Do You Smell What I Smell?			•			*Framework* DCIs: LS1.D, PS4.C *NGSS* PE: 1-LS1-1
36; February 2010	Helper Hats/Will It Protect Me From the Water?			•			*Framework* DCI: ETS1.B *NGSS* PEs: K-2-ETS1-2, K-2-ETS1-3
Water Play							
15; January 2006	The Matter of Melting/Melt Away!	•	•				*Framework* DCI: PS1.A *NGSS* PEs: 2-PS1-1, 2-PS1-4, 2-ESS2-3
24; July 2007	Water Works/How Can We Move Water?	•	•		•		*Framework* DCIs: PS2.A, ETS1.A, ETS1.B *NGSS* PEs: K-PS2, K-2-ETS1-3

Table 1.1 (continued)

Chapter number; original publication date	Title of column entry/Activity	Area of Science					Framework and NGSS connections
		Physical sciences*	Chemical sciences†	Life sciences	Earth and space sciences	NOS and nature of scientific inquiry	
Water Play							
33; July 2009	Adding Up the Rain/Measuring Up				•		*Framework* DCI: ESS2.D *NGSS* PE: K-ESS2-1
36; February 2010	Helper Hats/Will It Protect Me From the Water?			•			*Framework* DCI: ETS1.B *NGSS* PEs: K-2-ETS1-2, K-2-ETS1-3
37; March 2010	Building With Sand/How High Can You Build With Sand?				•	•	*Framework* Practice: Constructing explanations and designing solutions DCIs: PS1.A, ETS1.C *NGSS* PEs: 2-PS1-2, K-2-ETS1-3
39; October 2010	Developing Observation Skills/Making and Observing Bubbles	•				•	*Framework* Practices: Asking questions and defining problems; planning and carrying out investigations *NGSS* PE: 2-PS1-1
40; December 2010	Investigable Questions/How Much Is Enough?			•		•	*Framework* Practices: Asking questions; planning and carrying out investigations; analyzing and interpreting data *NGSS* PEs: K-LS1-1, 2-LS2-1

Table 1.1 (continued)

Chapter number; original publication date	Title of column entry/Activity	Physical sciences*	Chemical sciences†	Life sciences	Earth and space sciences	NOS and nature of scientific inquiry	Framework and NGSS connections
		Area of Science					
Water Play							
41; February 2011	Ongoing Inquiry/Properties of Water	•					*Framework* All practices
52; April 2013	Water Leaves "Footprints"/Sand Movers	•	•		•	•	*Framework* All practices *NGSS* PE: 2-ESS1-1
56; July 2014	Inviting Parents to School/Making Lemonade From Scratch		•				*Framework* DCI: PS1.A *NGSS* PEs: 2-PS1-1, 2-PS1-2
Weather							
15; January 2006	The Matter of Melting/Melt Away!	•	•				*Framework* DCI: PS1.A *NGSS* PEs: 2-PS1-1, 2-PS1-4, 2-ESS2-3
22; March 2007	The Sun's Energy/Making Sun Prints				•		*Framework* DCI: PS3.B *NGSS* PE: K-PS3-1
29; December 2008	Air Is Not Nothing/Moving Air		•				*Framework* DCI: PS1.A *NGSS* PEs: K-PS2-1, 1-PS4-1

Table 1.1 (continued)

Chapter number; original publication date	Title of column entry/Activity	Area of Science						Framework and NGSS connections
		Physical sciences*	Chemical sciences†	Life sciences	Earth and space sciences	NOS and nature of scientific inquiry		

Weather

33; July 2009	Adding Up the Rain/ Measuring Up				•			*Framework* DCI: ESS2.D *NGSS* PE: K-ESS2-1
41; February 2011	Ongoing Inquiry/ Properties of Water	•						*Framework* All practices
43; July 2011	Measuring Learning/ Temperature Changes	•						*Framework* Practice: Using mathematics and computational thinking *NGSS* PE: K-ESS2-1
50; January 2013	The Wonders of Weather/Observing Weather				•			*Framework* DCI: ESS2.D *NGSS* PE: K-ESS2

Designing and Building to Solve a Problem

17; July 2006	Roll With It/Wheel Work	•						*Framework* Practices: Planning and carrying out investigations; constructing explanations (for science) and designing solutions (for engineering). CCC: Structure and function *NGSS* PEs: K-PS2, K-2-ETS1-2

Table 1.1 (continued)

Designing and Building to Solve a Problem

Chapter number; original publication date	Title of column entry/Activity	Area of Science					Framework and NGSS connections
		Physical sciences*	Chemical sciences†	Life sciences	Earth and space sciences	NOS and nature of scientific inquiry	
24; July 2007	Water Works/How Can We Move Water?	•	•		•		*Framework* DCIs: PS2.A, ETS1.A, ETS1.B *NGSS* PEs: K-PS2, K-2-ETS1-3
27; March 2008	Objects in Motion/ Crash Dummy Fun!	•					*Framework* DCI: PS2.A *NGSS* PEs: K-PS2-1, K-PS2-2, K-2-ETS1-3
36; February 2010	Helper Hats/Will It Protect Me From the Water?			•			*Framework* DCI: ETS1.B *NGSS* PEs: K-2-ETS1-2, K-2-ETS1-3
37; March 2010	Building With Sand/ How High Can You Build With Sand?				•	•	*Framework* Practice: Constructing explanations and designing solutions DCIs: PS1.A, ETS1.C *NGSS* PEs: K-2-ETS1-3, 2-PS1-2
48; November 2012	Drawing Movement/ Just Draw It—Ramps	•					*Framework* Practices: Planning and carrying out investigations; designing solutions *NGSS* PEs: K-PS2-1, K-PS2-2

Table 1.1 (continued)

Chapter number; original publication date	Title of column entry/Activity	Area of Science					Framework and NGSS connections
		Physical sciences*	Chemical sciences†	Life sciences	Earth and space sciences	NOS and nature of scientific inquiry	
Designing and Building to Solve a Problem							
53; October 2013	The STEM of Inquiry/Heavy Lifting	•				•	*Framework* DCI: ETS1 *NGSS* PE: K-2-ETS1
54; November 2013	Are They Getting It?/Materials Have Properties	•				•	*Framework* DCIs: PS1.A, ETS1 *NGSS* PEs: 2-PS1, K-2-ETS1
Technology							
14; September 2005	What Can Young Children Do as Scientists?/Scientists at Work					•	*Framework* All practices *NGSS* Appendix H (Science is a human endeavor)
17; July 2006	Roll With It/Wheel Work	•					*Framework* Practices: Planning and carrying out investigations; constructing explanations (for science) and designing solutions (for engineering) CCC: Structure and function *NGSS* PEs: K-PS2, K-2-ETS1-2

Table 1.1 (continued)

Chapter number; original publication date	Title of column entry/Activity	Area of Science					Framework and NGSS connections
		Physical sciences*	Chemical sciences†	Life sciences	Earth and space sciences	NOS and nature of scientific inquiry	
Technology							
19; October 2006	Learning Measurement/Measuring Hands					•	*Framework* Practice: Using mathematics and computational thinking *NGSS* Appendix H (Scientific investigations use a variety of methods)
24; July 2007	Water Works/How Can We Move Water?	•	•		•		*Framework* DCIs: PS2.A, ETS1.A, ETS1.B *NGSS* PEs: K-PS2, K-2-ETS1-3
26; February 2008	Observing With Magnifiers/Exploring Magnifiers	•				•	*Framework* DCI: PS4.B *NGSS* PE: K-2-ETS1-2
31; March 2009	Does Light Go Through It?/Playing With Light	•					*Framework* DCI: PS4.B *NGSS* PEs: 1-PS4-3, 2-PS1-1
33; July 2009	Adding Up the Rain/Measuring Up				•		*Framework* DCI: ESS2.D *NGSS* PE: K-ESS2-1

Table 1.1 (continued)

Chapter number; original publication date	Title of column entry/Activity	Area of Science					Framework and NGSS connections
		Physical sciences*	Chemical sciences†	Life sciences	Earth and space sciences	NOS and nature of scientific inquiry	
Technology							
41; February 2011	Ongoing Inquiry/ Properties of Water	•					*Framework* All practices
42; April 2011	Sharing Research Results/Playdough Guidebook		•			•	*Framework* DCIs: PS1.A, PS1.B *NGSS* PEs: 2-PS1-1, 2-PS1-2, 2-PS1-4
43; July 2011	Measuring Learning/ Temperature Changes	•					*Framework* Practice: Using mathematics and computational thinking *NGSS* PE: K-ESS2-1
44; September 2011	A Sense of Place/ Schoolyard as Model				•		*Framework* Practice: Developing and using models *NGSS* PE: 2-ESS2-2
49; December 2012	Please Touch Museum/Materials Museum	•				•	*Framework* DCI: PS1.A *NGSS* PE: 2-PS1-1

Table 1.1 (continued)

Chapter number; original publication date	Title of column entry/Activity	Area of Science					Framework and NGSS connections
		Physical sciences*	Chemical sciences†	Life sciences	Earth and space sciences	NOS and nature of scientific inquiry	
Technology							
50; January 2013	The Wonders of Weather/Observing Weather				•		*Framework* DCI: ESS2.D *NGSS* PE: K-ESS2

Note: CCC = crosscutting concept; DCI = disciplinary core idea; NOS = nature of science; PE = performance expectation. The abbreviations in the DCIs and PEs refer to the grade levels (for purposes of this table, the level may be K, 1, 2, or K–2) and disciplinary areas (ESS = Earth and space sciences; ETS = engineering, technology, and applications of science; LS = life sciences; PS = physical sciences).

*Includes the concepts of force, motion, and light.

†Includes properties of matter, interactions, and chemical processes.

‡Science and engineering practices, CCCs, and DCIs are from *A Framework for K–12 Science Education* (National Research Council 2012); *NGSS* refers to the *Next Generation Science Standards* (NGSS Lead States 2013), including specific PEs and Appendix H ("Understanding the Scientific Enterprise: The Nature of Science in the *Next Generation Science Standards*").

Chapter 2

Assessment

A *Framework for K–12 Science Education* (*Framework*; NRC 2012) and the National Science Teachers Association Early Childhood Science Education position statement (see Appendix 2, pp. 347–352) recognize that young children are developmentally ready to build on and revise their knowledge and abilities in science and engineering. To help children revise their understanding and build their ability, teachers must first know what their students understand.

> *Assessment of young children is crucial in meeting a variety of purposes. It provides information with which caregivers and teachers can better understand individual children's developmental progress and status and how well they are learning, and it can inform caregiving, instruction, and provision of needed services. It helps early childhood program staff determine how well they are meeting their objectives for the children they serve, and it informs program design and implementation. It provides some of the information needed for program accountability and contributes to advancing knowledge of child development. (Committee on Developmental Outcomes and Assessments for Young Children 2008, p. 18)*

In 2003, the National Association for the Education of Young Children (NAEYC) and the National Association of Early Childhood Specialists in State Departments of Education issued a joint position statement on early childhood curriculum, assessment, and program evaluation. According to this statement, assessments in early childhood must be used for the specific and beneficial purposes of "(1) making sound decisions about teaching and learning, (2) identifying significant concerns that may require focused intervention for individual children, and (3) helping programs improve their educational and developmental interventions."

We assess children's learning to understand how to make our teaching more effective. Teachers design activities to develop and support science inquiry and engineering design based on what the children already know and what we want them to learn. The NAEYC position statement on developmentally appropriate practice states that in planning learning experiences, an effective teacher considers the ages of the children as well as the children's family, community, culture, linguistic norms, social group, past experience (including learning and behavior), and current circumstances (NAEYC 2009). Assessment may be made with teacher observations or by using a tool that has been validated for reliability, gathering evidence of children's development or learning (McAffee, Leong, and Bodrova 2004). Assessment of children in early childhood classrooms "is challenging because they develop and learn in ways that are characteristically uneven and embedded within the specific cultural and linguistic contexts in which they live" (NAEYC 2009, p. 22).

Through assessment we can gather information about the need for accommodations and services for successful learning. Conversation and discussion are important in assessment—during individual conversations and group discussions children often reveal their ability to analyze and their understanding of concepts. Documenting assessments we make through observation helps us keep track of what we learn about children's understanding. We can share the knowledge about children that we gain through assessment with colleagues and families—for example, which children are especially interested in living organisms, how does a child reconcile a new experience with a previously held contradictory idea, and which children need more support to follow through on a difficult task.

Plan ahead of time how you will find out what the children have learned thus far from their participation in science inquiry. By being prepared to collect evidence of thinking, ability, and understanding, you will be able to gather the information needed to know when to offer additional materials or challenges. If we know children's understanding of a concept, such as "water is a liquid," we can make decisions about further activities and information to use to engage children in learning more about matter, such as, "Water is a solid when it is frozen." It is equally important to know children's ability to ask questions, persist in solving a problem, try alternatives, discuss ideas, use mathematics, use evidence to support their ideas, and try to find answers to their questions. These are some of the science and engineering practices (NRC 2012) that support the science education goal of participating in inquiry, beginning in early childhood as we encourage, not dismiss, children's questions and consider how we can help children stretch to reach new understanding and gain new skills to investigate the world.

The curriculum you use may support a particular assessment format. There are research-based early childhood science learning assessment tools with a systematic approach that have reliability and validity data (Brenneman 2011). Table 2.1 (pp. 44–48) shows an example of an assessment tool that I developed; the table includes sequences of children's science behavior in early childhood, with related science and engineering practices identified in the *Framework* (NRC 2012). In developing this tool, I was influenced and inspired by discussions and examples of assessments in the *Framework* as well as in Paysnick (2010) and the *Young Scientist Series* (Chalufour and Worth 2003–2005).

The assessment tool shown in Table 2.1 lists child behaviors I look for to determine if the child is learning how to participate in science inquiry from the experiences. This tool has not been validated for reliability, nor is it part of a larger system of educational services, even though both are important factors to consider when choosing assessment tools. Nevertheless, I use it to see if I am keeping up with the children's needs for discussion of ideas and introduction of new materials. I offer this tool as an example of how a teacher might closely examine student behavior to assess science and engineering learning. Note that the listed behaviors fall within the early childhood age range, but teachers will need to adapt this tool to particular classes and programs. Learning science and engineering content—knowledge about the world and designing solutions to problems—is important, but, because the set of information that students learn will vary depending on the inquiry they follow, I don't offer a content assessment here. Children who learn how to use science and engineering practices to investigate and design solutions can apply their reasoning and use of practices to acquire knowledge.

References

Brenneman, K. 2011. Assessment for preschool science learning and learning environments. *ECRP: Early Childhood Research and Practice* 13 (1). *http://ecrp.uiuc.edu/v13n1/brenneman.html*.

Chalufour, I., and K. Worth. 2003–2005. Young Scientist Series. St. Paul, MN: Redleaf Press.

Committee on Developmental Outcomes and Assessments for Young Children; Snow, C. E., and S. B. Van Hemel, eds. 2008. *Early childhood assessment: Why, what, and how.* Washington, DC: National Academies Press. *www.nap.edu/catalog/12446/early-childhood-assessment-why-what-and-how.*

McAffee, O., D. J. Leong, and E. Bodrova. 2004. *Basics of assessment: A primer for early childhood educators.* Washington, DC: National Association for the Education of Young Children.

National Association for the Education of Young Children (NAEYC). 2009. *Position statement: Developmentally appropriate practice in early childhood programs serving children from birth through age 8.* Washington, DC: NAEYC. *www.naeyc.org/files/naeyc/file/positions/PSDAP.pdf.*

National Association for the Education of Young Children (NAEYC) and the National Association of Early Childhood Specialists in State Departments of Education (NAECS/SDE). 2003. *Early childhood curriculum, assessment, and program evaluation.* Washington, DC: NAEYC and NAECS/SDE. *www.naeyc.org/files/naeyc/file/positions/pscape.pdf.*

National Research Council (NRC). 2012. *A framework for K–12 science education: Practices, crosscutting concepts, and core ideas.* Washington, DC: National Academies Press. *www.nap.edu/catalog/13165/a-framework-for-k-12-science-education-practices-crosscutting-concepts.*

Paysnick, R. 2010. Finding learning beneath the surface: Monitoring student progress with Science Practice Learning Progressions. *Science Scope* 34 (2): 69–76.

Additional Resource

Riley-Ayers, S. 2014. An assessment primer: What is effective assessment in the early childhood classroom? Presentation made at a meeting of the San Antonio Association for the Education of Young Children. *http://nieer.org/sites/nieer/files/Assessment%20presentation%20San%20Antonio%20AEYC%202014.pdf.*

TABLE 2.1. Example of an assessment tool showing sequences of children's science behavior in early childhood, with related scientific and engineering practices

Child behavior	Levels	Related scientific and engineering practices*
Child asks questions, initiates exploration, makes a prediction or guess, and tests a prediction, and may design and carry out impromptu investigations. *Examples:* • A child asks, "How can I build a house?" and begins stacking blocks. • A child points to a block tower being built by another child and says, "It's going to fall!" Later the child may say, "I'm building a tower that is taller and it won't fall."	• Child asks questions. • Child asks questions, initiates exploration. • Child asks questions, initiates exploration, makes a prediction or guess, and tests a prediction without preplanning. • Child asks questions, initiates exploration, makes a prediction or guess, and designs and carries out an impromptu investigation to test the prediction.	• Asking questions (for science) and defining problems (for engineering) • Planning and carrying out investigations
Child shows that he or she observes with purpose to find answers to questions. *Examples:* • If a child asks, "What is the caterpillar doing?" she looks closely to find out what the caterpillar is doing (eating, crawling, or other activity). • A child says, "I'm going to look tomorrow to see if the caterpillar ate the leaf."	• Child uses his or her sense after a question is asked. • Child uses more than one sense after a question is asked. • Child expresses a purpose for using more than one sense to find answers to the question. • Child describes how he or she will observe to find answers to the question.	• Planning and carrying out investigations
Child draws and builds models of things. *Examples:* • A child uses a toy dump truck to carry sand. • A child holding a block scrolls using a finger to pretend to show a photo on the "phone." • Using playdough, a child constructs a bird nest with eggs.	• Child draws simple pictures of objects and uses models of things such as pretend food and toy cars. • Child draws pictures of things with parts and makes models from simple parts, such as a few interlocking blocks or sticks. • Child draws pictures of things with parts and then constructs the model.	• Developing and using models • Obtaining, evaluating, and communicating information

Table 2.1 (*continued*)

Child behavior	Levels	Related scientific and engineering practices*
Child helps construct graph by adding data in appropriate places and compares amounts on graph—less or more. *Examples:* • A child says, "I have a red apple" and puts the apple in the line with the other red apples. • Children count the number of cloudy day symbols and the number of sunny day symbols on the calendar and compare the amounts. They discuss whether the data relate to the current season.	• Child tells the recorder in which group to put his or her data. Child counts as recorder points to units on graph. • Child marks or places data in appropriate group on graph. Child points to graph and counts. • Child records data accurately on graph and compares groups on the graph. • Child records data accurately on graph, compares groups on the graph, and discusses what the data comparison means.	• Analyzing and interpreting data • Using mathematics and computational thinking
Child uses tools, such as magnifiers and measuring tools, purposefully. *Examples:* • A child holds a magnifier up to his eye. He comments on how the magnifier "makes it bigger." • A child uses interlocking cubes to count how tall a plant has grown.	• Child attempts to use tool for its purpose when it is introduced. • Child works with tool for best use when it is given to him or her. • Child declares need for a tool to view or measure an object and is successful in using tool for the designed use. • Child gets the needed tool and successfully uses it as designed to view or measure object.	• Using mathematics and computational thinking
Child understands that measurement is a comparison between objects or sets. *Examples:* • Lifting a relatively big rock, a child says, "It's really heavy." • With a length of ribbon, a child wraps it around a container and says, "It goes all the way around." • Watching a caterpillar crawl on a ruler, a child says, "It's more than one inch long."	• Child describes the size or amount of an object or set with words or gestures. • Child compares an object or set to another known object or set. • Child uses a measurement tool and describes an object in comparison to the tool. • Child uses a measurement tool as designed to determine an amount that describes the object.	• Using mathematics and computational thinking • Obtaining, evaluating, and communicating information

Table 2.1 (*continued*)

Child behavior	Levels	Related scientific and engineering practices*
Child builds to solve a problem and draws to design a solution. *Example:* • A ramp built by children using several lengths of blocks keeps falling down and the children put it back up. Later a child says the structure needs more blocks to better support the ramp section, and the children add some. In a group discussion, the children draw a "strong" ramp structure together showing many supporting blocks.	• Child experiences a difficulty with materials during play and makes an adjustment to the materials to try to solve the difficulty. • Child experiences a difficulty with materials during play and makes adjustments until difficulty is overcome. • Child participates in a discussion about a difficulty with materials and talks about possible solutions. • Child participates in a discussion about a difficulty with materials, talks about or draws possible solutions, and makes adjustments to overcome a difficulty with materials.	• Constructing explanations (for science) and designing solutions (for engineering)
Child discusses observations and explains his or her thinking. *Examples:* • A child touches a soft blanket and then extends the contact with it by hugging it. • Hearing a wind chime, a child says, "Is the wind making that sound?" "A second child says, "Yes, the wind blows in them," and another child says, "No, the pieces make that sound when they get blown together."	• Child makes an observation about what he or she senses and communicates by gesture or words. • Child tells, writes, or draws about what he or she sees or otherwise senses and expresses a wondering. • Child discusses the observation with others, explains it, and expresses a wondering. • Child discusses the observation, proposes an explanation, and explains his or her thinking.	• Constructing explanations (for science) and designing solutions (for engineering) • Engaging in argument from evidence • Obtaining, evaluating, and communicating information

Table 2.1 (*continued*)

Child behavior	Levels	Related scientific and engineering practices*
Child counts and compares the data, and describes his or her changes in thinking, if any. *Example:* • A child counts the number of scoops of water that fit into a container. Using a bigger scoop, she counts again, compares the number of small scoops with the number of large scoops, and says, "You need more small scoops to fill the container."	• Child counts for the purpose of collecting data. • Child counts for the purpose of collecting data and compares the count with another set. • Child counts for the purpose of collecting data, compares the count with another set and reflects on the meaning of the comparison. • Child counts for the purpose of collecting data, compares the data, reflects on the meaning of the comparison, and describes his or her changes in thinking, if any.	• Using mathematics and computational thinking • Engaging in argument from evidence
Child discusses and/or draws observations or ideas. *Example:* • A child points to a cricket and says, "It's moving!" After drawing the cricket, he says, "It has so many legs to jump high with."	• Child says, "Look!" and/or points to object. • Child talks about the object observed. • Child draws observation and tells what it is. • Child draws observation and tells what it means.	• Engaging in argument from evidence • Obtaining, evaluating, and communicating information
Child shows that he or she is observing with more than one sense, by focus, gesture, drawing, or written or spoken word. *Example:* • A child handles a musical triangle. She uses the striker and makes a sound and says, "It tickles (vibrates)." By varying the strength of the tap, the child makes the sound go from soft to loud. She says, "I made it get louder!"	• Child looks intently at, or listens intently to, an object. • Child looks, touches, smells, and/or listens to an object. • Child communicates one detail of what he or she sees, feels, smells or hears. • Child communicates two or more details of what he or she sees, feels, smells, or hears.	• Obtaining, evaluating, and communicating information

Table 2.1 (*continued*)

Child behavior	Levels	Related scientific and engineering practices*
Child seeks out information from resources such as other people, books, and posters. *Example:* • Looking in a mirror, a child asks, "How many teeth do I have?" Looking in books, he counts the number of teeth in a diagram of a child's mouth and looks for teeth in another book of animals.	• Child asks for more information. • Child asks where to get more information. • Child looks in book to find information in text or pictures. • Child seeks out information from multiple resources such as other people, books, and posters.	• Obtaining, evaluating, and communicating information

Note: The levels suggest steps in a child's growing ability but do not correspond to specific ages. The examples show only a few of the many possible situations where the behavior may be seen.

*The following scientific and engineering practices for K–12 classrooms are identified in *A Framework for K–12 Science Education: Practices, Crosscutting Concepts, and Core Ideas* (NRC 2012):

- Asking questions (for science) and defining problems (for engineering)
- Developing and using models
- Planning and carrying out investigations
- Analyzing and interpreting data
- Using mathematics and computational thinking
- Constructing explanations (for science) and designing solutions (for engineering)
- Engaging in argument from evidence
- Obtaining, evaluating, and communicating information

Part II.

Introduction to the Topics of *The Early Years Column Entries*

Chapter 3

Nature of Science and Science Process Skills

Science is a process based on observations of the natural world. It is also a body of knowledge and a way of knowing or building knowledge about the world. The nature of science (NOS) is usually described as having six or more *basic understandings* (NGSS Lead States 2013), also called *aspects* by many researchers (Crowther, Lederman, and Lederman 2005; Lederman and Lederman 2004; Lederman et al. 2014; NSTA 2000; Quigley, Buck, and Akerson 2011):

- Science addresses questions about the natural and material world.

- Science is based on and/or derived from observations of the natural world.

- Scientific knowledge is based on a variety of methods.

- Science requires understanding the difference between observations and inferences.

- Scientific knowledge is subject to change as new information or understanding based on evidence develops.

- Science is subjective, because scientific work is done by people who have biases and mind-sets that can influence understanding.

- Science is creative, because scientists think creatively to make inferences and explanations.

- Science is affected by the social and cultural structure.

The Understanding Science website of the University of California Museum of Paleontology states that "Kindergarten, first, and second grade students can begin to understand what science is, who does science, and how scientists work through classroom activities, stories about scientists, and class discussions" (Janulaw n.d.) Although this website says that investigating questions, gathering evidence through observations, and holding class discussions to share evidence and

ideas are appropriate activities for first grade, beginning efforts in these tasks can start in preschool with developmentally appropriate expectations.

Most of *The Early Years* column entries include some attention to the NOS. Beginning with providing many opportunities to make observations—describe objects and interesting happenings—early childhood educators can guide children as they build an understanding of the NOS. Scientists make inferences or best possible explanations based on observations, their prior knowledge, and their ability to creatively imagine possibilities. Children can learn the difference between observation and inference somewhere around the age of 5. Plan how to make the aspects of the NOS visible to children as you plan the curriculum. Here is a very simple example of how to voice your observations, make inferences and offer evidence, and model how understanding can change with new information:

> *Make observations: "I hear a sound outside."*

> *Make inferences and offer evidence: "I think it is raining because that sound is like the sound I heard when it rained yesterday."*

> *Model how your understanding can change with new information: "I looked out the window and saw the sprinkler spraying water. It wasn't what I thought was happening."*

Another opportunity to include NOS discussion in science activities can happen when children are making inferences about what is inside a closed container, such as in Chapter 18, "Young Questioners." Children often say that they think their favorite object, a red ball, is the unseen object inside their container, even when the evidence of sound and weight do not suggest that inference. They are biased toward having the red ball in their container because that is what they want. Teachers can model this bias by saying, "I think I have the red ball in my container because even though it sounds like a coin and is not heavy, I want it to be the red ball." After opening the container to check, we can model our understanding of the bias, "I wanted the ball but I heard a sound like a coin and the container wasn't heavy. The evidence was not for the ball, but I said what I wanted, not what the evidence pointed to." Teachers can also model how their understanding of a topic can change as they make new observations or get new information. For example, one might say, "I thought all metal objects were attracted by magnets, but when I tested these metal keys they were not attracted by the magnet! Some metal objects are not attracted by magnets."

These small opportunities can build children's understanding over time. Look for opportunities to make statements about the NOS.

References

Crowther, D. T., N. G. Lederman, and J. S. Lederman. 2005. Methods and strategies: Understanding the true meaning of nature of science. *Science and Children* 43 (2): 50–52.

Janulaw, S. n.d. Know your students: Implications for understanding the nature of science. University of California Museum of Paleontology. *http://undsci.berkeley.edu/teaching/k2_implications2.php.*

Lederman, J., S. Bartels, N. Lederman, and D. Gnanakkan. 2014. Demystifying nature of science: Two activities help young children understand nature of science. *Science and Children* 52 (1): 40–45.

Lederman, N. G., and J. S. Lederman. 2004. Revising instruction to teach nature of science. *The Science Teacher* 71 (9): 36–39.

National Science Teachers Association (NSTA). 2000. NSTA position statement: The nature of science. *www.nsta.org/about/positions/natureofscience.aspx.*

NGSS Lead States. 2013. *Next Generation Science Standards: For states, by states.* Washington, DC: National Academies Press. *www.nextgenscience.org/next-generation-science-standards.*

Quigley, C., G. Buck, and V. Akerson. 2011. The nature of science challenge. *Science and Children* 49 (2): 57–61.

Additional Resources

Akerson, V., G. Buck, L. Donnelly, V. Nargund-Joshi, and I. Weiland. 2011. The importance of teaching and learning nature of science in the early childhood years. *Journal of Science Education and Technology* 20 (5): 537–549.

Edson, M. T. 2013. *Starting with science.* Portland, ME: Stenhouse.

Chapter 4

Objects in Motion

Parents, early childhood teachers, and education researchers have noted that children begin learning physical science concepts as babies by dropping a spoon from the high chair, over and over. This early investigation into how solid objects move ("the simple mechanics of solid bounded objects" [Michaels, Shouse, and Schweingruber 2008, p. 38]) meets the criteria of researchers for an age-appropriate science activity (Worth 2010):

- The phenomena selected for young children must be available for direct exploration and drawn from the environment in which they live.

- The concepts underlying the children's work must be concepts that are important to science.

- The focus of science must be on concepts that are developmentally appropriate and can be explored from multiple perspectives, in depth, and over time.

- The phenomena, concepts, and topics must be engaging and interesting to the children and their teachers.

Physical science concepts about motion that are commonly part of early childhood standards include "pushes and pulls can have different strengths and directions, and can change the speed or direction of its motion or start or stop it" (NGSS Lead States 2013) and describing the effects of magnets (Virginia Department of Education, Office of Humanities and Early Childhood 2013). Children describe pushes and pulls as big or small forces, building an understanding of comparison and measurement that will become more precise in middle and high school using standard measurements.

This topic section (see Table 1.1, pp. 15–39) includes early childhood activities for ages 3 and up that direct children's exploration to the concepts of force, motion, and the properties of matter. The activities are not intended to stand alone but should be incorporated into a science inquiry of a single concept over a period of weeks or months. Preschool-age children have developed their thinking about the world to form beginning theories, including an understanding of the principles of cause and effect (Michaels, Shouse, and Schweingruber 2008, p. 40). This doesn't mean that they are ready to understand high school physics vocabulary such as inertia and acceleration, but they are ready to continue their explorations into how objects behave in motion and into pushes and pulls.

Early childhood teachers are also ready—ready to support science inquiry explorations by learning a little bit more about the science through their own explorations, and by reading books such as this one and books for early elementary students on specific science concepts. Nonfiction science books for grades 2–4 can explain science concepts at a level just above kindergarten so teachers who are also beginning to learn concepts and vocabulary can learn what is appropriate for early childhood. Teacher-only explorations are important for understanding what challenges in thinking and learning might crop up for the children as they use the materials. By using the same materials used by the children, teachers can experience the insights that are gained when people are able to make their own discoveries and solve their own problems, something that teachers will then allow children to do. It is also helpful to do activities with other teachers (perhaps at a meeting to plan curriculum) so you will have discussion partners to share ideas about your observations and plan next steps for classroom science learning.

Some ongoing physical science explorations (such as working with ramps and objects to roll down them) are easy to implement because the materials are commonly available, the setup can be done by the children, the space requirements are flexible, they do not require daily attention or care, and the cleanup can also be completed by the children. Keep in mind that you do not have to present many different science activities for your students to learn about the nature of science and develop the skills needed to participate in science inquiry. Spending a month or more investigating how water moves leads to deeper understanding than a series of brief, unrelated activities such as a day each on investigating ramps, light and shadow, plant growth, and magnetic force. Children need lengths of uninterrupted time to explore the materials and begin to develop their ideas and solve problems. A period of 45–60 minutes is not too long—extended periods of time allow children

to become deeply involved with the materials and exploring the concepts (Copple and Bredekamp 2009).

Beginning with a simple activity of handling, describing, and rolling a variety of three-dimensional shapes, children can become familiar with how the form of an object affects its motion (see Chapter 17, "Roll With It"). As they have additional experiences moving and dropping objects and building structures to roll objects on, children have firsthand experience with the effects of pulls and pushes they apply and those occurring in nature (gravity and friction) (see Chapters 27, "Objects in Motion," p. 160; 46, "Seeing the Moon," p. 271; and 48, "Drawing Movement," p. 282). Participating in these activities, used in any sequence, and talking about their observations, support children's developing ideas about force and motion and are foundational for later learning in physics classes.

Other physical science activities involve the states of matter: gas, liquids, and solids (see Chapters 24, "Water Works," p. 141; 29, "Air Is Not Nothing," p. 171; 32, "Hear That?" p. 190; 41, "Ongoing Inquiry," p. 243; and 43, "Measuring Learning," p. 255). Having experiences with materials and explaining their properties and any changes they see prepare children for future learning about atomic structure and energy transfer.

References

Copple, C., and S. Bredekamp, eds. 2009. *Developmentally appropriate practice in early childhood programs serving children from birth through age 8.* 3rd ed. Washington, DC: National Association for the Education of Young Children.

Michaels, S., A. W. Shouse, and H. Schweingruber. 2008. *Ready, set, SCIENCE! Putting research to work in K–8 science classrooms.* Washington, DC: National Academies Press.

NGSS Lead States. 2013. Appendix E—Progressions within the *Next Generation Science Standards.* In *Next Generation Science Standards: For states, by states.* Washington, DC: National Academies Press. *www.nextgenscience.org/next-generation-science-standards.*

Virginia Department of Education, Office of Humanities and Early Childhood. 2013. *Virginia's foundation blocks for early learning: Comprehensive standards for four-year-olds. www.doe.virginia.gov/instruction/early_childhood/preschool_initiative/foundationblocks.pdf.*

Worth, K. 2010. Science in early childhood classrooms: Content and process. Paper presented at the SEED (STEM in Early Education and Development) conference at the University of Northern Iowa. *http://ecrp.uiuc.edu/beyond/seed/worth.html.*

Additional Resources for Physical Science

Keeley, P., and R. Harrington. 2010. *Uncovering student ideas in physical science, vol. 1: 45 new force and motion assessment probes.* Arlington, VA: NSTA Press.

Robertson, W. 2002. *Force and motion: Stop faking it! Finally understanding science so you can teach it.* Arlington, VA: NSTA Press.

National Science Teachers Association

Chapter 5

What Are Things Made of?

Every day, children explore the states of matter (solid, liquid, and gas) and the properties of materials (soft, hard, squishy, blue, rough, big, wet, cold, etc.). Sometimes the exploration is part of answering a question, such as, "Where does the water go when I pour it into the sand-box?" or "Can a rock melt in the Sun?" Early experiences with materials prepare children to make conceptual changes to incorporate new understandings in later years as they begin to understand that air is matter, too, and that matter is something that has mass and takes up space (Michaels, Shouse, and Schweingruber 2008, p. 42). As described in several state and national standards (Massachusetts Executive Office of Education 2013; NGSS Lead States 2013; Virginia Department of Education, Office of Humanities and Early Childhood 2013), the following are early childhood concepts about matter:

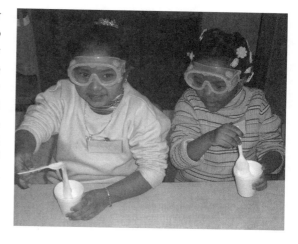

- There are many different kinds of matter, called substances. Substances can exist in liquid and solid forms.

- Matter has properties that can be observed.

- Mixing different substances together can make a change.

- Matter can exist in natural or human-manufactured forms.

The properties of materials must be considered when designing a solution to a problem. Young architects and engineers take the properties of materials into consideration as they build something, perhaps a bridge for toy cars using cardboard, sticks, and tape. Take advantage of the rich vocabulary we use to describe materials, and include writing and learning new vocabulary as part of your science and engineering teaching. Children use mathematics in service to investigating properties that can be measured and quantified—for example, how sticky, warm, or heavy is the material?

The activities in this topic section (see Table 1.1, pp. 15–39) use light to explore matter as it moves through magnifying lenses and other materials, investigate processes including melting and dissolving, and provide experiences that can help children understand the properties of water. See also Chapter 4 for a discussion of activities that will extend your students' investigation into properties of matter as they mix and make a change.

Experiences exploring the properties of matter with light can begin with the common activity of mixing colors. Adding activities using flashlights and other light sources to observe the direction of light and what happens when light is blocked deepens children's understanding of light—a phenomenon that is intriguing because it is observable but cannot be held. Early experiences with light are the foundation for later understanding about how light is used in technologies and how evidence of shadows helps astronomers reason about objects in space. As children explore what they can make happen with a light source, conversations with teachers about their experiences will help them put their ideas in words.

After introducing the words *liquid* and *solid* when you make playdough in the classroom, continue using them as you provide other experiences with liquids and solids and making changes to them (see Chapters 15, "The Matter of Melting," p. 88; 20, "Rocks Tell a Story," p. 117; 37, "Building With Sand," p. 220; 54, "Are They Getting It?" p. 322; and 56, "Inviting Parents to School," p. 335). Children's understanding of these concepts becomes much more solid when they have multiple experiences over time and can use the vocabulary in everyday conversation.

References

Massachusetts Executive Office of Education. 2013. *Pre-K science, technology and engineering standards. www.mass.gov/edu/birth-grade-12/early-education-and-care/curriculum-and-learning.*

Michaels, S., A. W. Shouse, and H. Schweingruber. 2008. *Ready, set, SCIENCE! Putting research to work in K–8 science classrooms.* Washington, DC: National Academies Press.

NGSS Lead States. 2013. Appendix E—Progressions within the *Next Generation Science Standards.* In *Next Generation Science Standards: For states, by states.* Washington, DC: National Academies Press. *www.nextgenscience.org/next-generation-science-standards.*

Virginia Department of Education, Office of Humanities and Early Childhood. 2013. *Virginia's foundation blocks for early learning: Comprehensive standards for four-year-olds. www.doe.virginia.gov/instruction/early_childhood/preschool_initiative/foundationblocks.pdf.*

Chapter 6

Mixing and Making a Change: Chemical Science

Mixing and making a change engages children in learning chemistry long before they are able to understand the atomic structure of materials. Explorations in early childhood introduce scientific core ideas that will build toward understanding atomic structure in high school. These ideas overlap with the ideas explored in activities described in Chapter 5, "What Are Things Made Of?" (p. 57):

- Matter exists as different substances that have observable different properties. Different properties are suited to different purposes. Objects can be built up from smaller parts (NGSS Lead States 2013).

- Heating and cooling substances cause changes that are sometimes reversible and sometimes not (NGSS Lead States 2013).

- It is possible to recognize through investigation that physical objects and materials can change under different circumstances (Massachusetts Executive Office of Education 2013).

Chemistry overlaps with biology and physics as a science that seeks to understand what materials are made of and how they combine in living and nonliving things. In early childhood, mixing materials together might happen in the kitchen, in the classroom, or on the playground. The cycles of inquiry that apply to understanding pushes and pulls in physics also apply to exploring materials while making playdough and exploring the properties of air: asking questions, making observations, recording data, making changes, asking more questions, talking about ideas, exploring additional materials, recording additional data, reflecting on data, trying something different, asking questions, and so on.

Activities involving liquids, or mixing materials together, take additional planning to ensure safety. Give children safety goggles to protect their eyes from splashes and dust, have tools to mop up spills to prevent falls, and be sure children wash their hands before and after the activity. Of course, you should follow up to be sure that all safety precautions are used.

Help children understand the difference between cooking activities and those you do as part of a science inquiry. This division is somewhat arbitrary in early childhood but is part of a crucial safety routine in school science laboratories. When making pancakes at home, children rarely use safety goggles, but they should wear them in science activities that involve making a mixture with small particles (e.g., sand, flour, or salt), which could get in their eyes. It is safe lab practice, and children enjoy wearing the goggles for short periods of time.

In cooking and chemistry explorations, children learn the importance of measuring volume and time and how to do it accurately. When children have the experience of making something that "didn't turn out right," they begin to understand the reason for measurement and the importance of following a procedure. What materials would you put in an activity center for open-ended exploration of making a mixture?

Like other activities described in the column entries related to this topic, Making Lemonade From Scratch (see Chapter 56, "Inviting Parents to School," p. 335) has children handling solids and liquids and combining them to make a change. Some of the changes described in these activities can be reversed and others cannot. If you begin an exploration of mixing with the lemonade activity, children will see the solid sugar grains "disappear" in the water. The dissolved sugar can be "found" again several days later by letting the water evaporate, leaving the sugar behind.

Children can mix other substances with water. Salt and sand are both safe and familiar to young children. Comparing the results of those mixtures with what happens to sugar helps children understand that some solids dissolve in water while others do not, a beginning understanding of the properties of matter. In the Ice Cream Science activity (see Chapter 55, "Now We're Cooking," p. 328), children will experience a change due to a change in temperature as well as a solid (sugar) dissolving in a liquid (cream). Continue exploring changes due to temperature with safe melting experiences (see Chapter 15, "The Matter of Melting," p. 88) and cooking playdough (see Chapter 42, "Sharing Research Results," p. 249).

There are additional articles about exploring mixing and making a change in the online archives of the National Science Teachers Association's journal *Science and Children;* one example explores the mixture of cornstarch and water (Ashbrook 2008). By allowing children to repeat experiences, as long as they are interested, we extend their understanding of the nature of science—that explorations can be done over and over—and the children can find out if the same changes happen to the matter every time.

References

Ashbrook, P. 2008. The early years: Exploring the properties of a mixture. *Science and Children* 45 (5): 18–20. Also available online at *www.nsta.org/store/product_detail. aspx?id=10.2505/4/sc08_045_05_18.*

Massachusetts Executive Office of Education. 2013. *Pre-K science, technology and engineering standards. www.mass.gov/edu/birth-grade-12/early-education-and-care/curriculum-and-learning.*

NGSS Lead States. 2013. Appendix E—Progressions within the *Next Generation Science Standards.* In *Next Generation Science Standards: For states, by states.* Washington, DC: National Academies Press. *www.nextgenscience.org/next-generation-science-standards.*

Chapter 7

Exploring Our Senses

Preschool curricula often begin in fall with an "All About Me" unit. Children enjoy exploring the world with the "tools" of their senses. Using our senses in an intentional way helps us find the limits of those senses and understand how technology might be useful in extending them. Activities about exploring our senses tie in easily with curricula about art, music, and healthy habits, becoming a richer inquiry. Children's vocabulary to explain sensory experiences can be clarified or expanded beyond *nice* and *soft* (Ashbrook 2007).

Early childhood science standards related to the senses include learning that all organisms have external parts that the organisms use to sense and communicate information (Massachusetts Executive Office of Education 2013; NGSS Lead States 2013; Virginia Department of Education, Office of Humanities and Early Childhood 2013). Beyond understanding that we have five senses and the related vocabulary, teachers can use sensory explorations to begin discussions about how our senses help us meet our needs to survive. Children may be interested in observing other animals to see if they have similar or different body parts for sensing the world.

Discussions about science concepts such as "Animals use their senses to survive" and engineering design problems such as "How can we make a tool to help us hear soft sounds?" may take place at any time in the school day. Documenting children's ideas and comments for later in-depth exploration supports children's thinking outside of "science time." Just as proper use of pronouns does not end when language arts time is over, early childhood investigations continue throughout out the day.

A simple game of "Stop and Listen" (see Chapter 32, "Hear That?" p. 190) can begin an investigation into hearing and the many questions that can be investigated, including what sounds can be heard from far away, how we can describe sounds, what the relationship is between the property of materials and the nature of sound, and whether we feel sounds. After engaging their senses in activities (Chapter 31, "Does Light Go Through It?" p. 184; Chapter 35, "Safe Smelling," p. 208; Chapter 49, "Please Touch Museum," p. 289; and Chapter 56, "Inviting Parents to School," p. 335), encourage children to explore the tools people use to extend our senses, connecting a senses investigation to use of technology (see

Chapter 26, "Observing With Magnifiers," p. 154). Discuss how the senses help animals survive and meet their needs (see Chapter 16, "Feet First," p. 93; see also Chapter 23, "Collards and Caterpillars," p. 135, in the "Learning About Plants" topic section [Table 1.1, pp. 15–39]).

References

Ashbrook, P. 2007. The early years: Recording sensory words. *Science and Children* 45 (4): 18–20.

Massachusetts Executive Office of Education. 2013. *Pre-K science, technology and engineering standards. www.mass.gov/edu/birth-grade-12/early-education-and-care/curriculum-and-learning.*

NGSS Lead States. 2013. Appendix E—Progressions within the *Next Generation Science Standards*. In *Next Generation Science Standards: For states, by states*. Washington, DC: National Academies Press. *www.nextgenscience.org/next-generation-science-standards.*

Virginia Department of Education, Office of Humanities and Early Childhood. 2013. *Virginia's foundation blocks for early learning: Comprehensive standards for four-year-olds*. Richmond, VA: Virginia Department of Education. *www.doe.virginia.gov/instruction/early_childhood/preschool_initiative/foundationblocks.pdf.*

Chapter 8

Learning About Plants and Animals: Biological Science

As described in several state and national early childhood standards (e.g., Massachusetts Executive Office of Education 2013; NGSS Lead States 2013; Virginia Department of Education, Office of Humanities and Early Childhood 2013), learning about living organisms often includes learning that

- living organisms have certain characteristics that distinguish them from nonliving objects—growth, movement, response to the environment, having offspring, and the need for food, air, and water;

- plants and animals change as they grow, have varied life cycles, and eventually die;

- the senses allow animals to react to their environment; and

- many young plants and animals are similar but not identical to their parents and to one another.

Children develop this understanding by taking care of living organisms and observing their changes as they grow and die. When the caretaking job of watering the seedlings or feeding the hamster is added to the job chart, it becomes important to the children and is noticed daily so the needs of the living things are met. Children notice more details of living organisms when they document and discuss any changes that occur as the organisms grow. Lengthening roots, blooming flowers, worms that crawl without legs, and caterpillars that eat, climb, and form chrysalides are all interesting to young children. They are also interested in death as part of the life cycle. Children often ask me about the dead insect they found: "What happened to it?" I tell them that all living things have life cycles that begin and end. The insect may have just gotten to the end of its life cycle, which is usually short in insects, or it may have had a sickness and could not get well.

Whether you introduce seed sprouting in the fall or in the spring, a science inquiry into the needs of living things can begin in the fall and continue all year. Student interest may wax and wane, but as you continue to offer new challenges and ask productive questions, they will re-engage with the process of understanding living organisms.

A child's favorite living organism might be him- or herself! Use the lessons in the topic section "Exploring Our Senses" (see Chapter 7) to begin discussion and investigation into the needs of living organisms. Asking your children, "Do you need air?" encourages them to think about this requirement for life, one that is often taken for granted. Chapter 40,

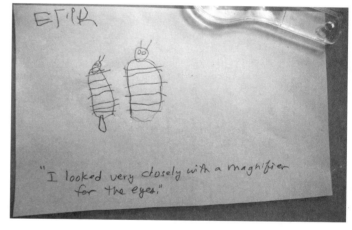

"I looked very closely with a magnifier for the eyes."

"Investigable Questions" (p. 237) raises the same question about plants. Other questions relate to parts of plants (Chapter 30, "Bring On Spring: Planting Peas," p. 177; Chapter 34, "Planting Before Winter," p. 202; and Chapter 35, "Safe Smelling," p. 208) and the relationship between plants and animals (Chapter 23, "Collards and Caterpillars," p. 135; Chapter 28, "An Invertebrate Garden," p. 165; and Chapter 51, "'Life' Science," p. 305). Making observations of other animals can reveal how they meet their needs for food and shelter and how their body structures change as they grow to adulthood (Chapter 21, "Birds in Winter," p. 124; Chapter 23, "Collards and Caterpillars," p. 135; and Chapter 25, "Counting a Culture of Mealworms," p. 147). Children can look for evidence to answer questions such as, "Where is the mommy?" "How many legs does a caterpillar have?" and "Does it have a mouth?" as they investigate living organisms.

An inquiry into living organisms would not be complete without an exploration and discussion about the role of sunshine in meeting their needs. The vital role of solar radiation to maintaining life on Earth is beyond most young children, but they can begin to sense it and observe the effects of the Sun's rays (see Chapter 22, "The Sun's Energy," p. 130).

References

Massachusetts Executive Office of Education. 2013. *Pre-K science, technology and engineering standards. www.mass.gov/edu/birth-grade-12/early-education-and-care/curriculum-and-learning.*

NGSS Lead States. 2013. Appendix E—Progressions within the *Next Generation Science Standards*. In *Next Generation Science Standards: For states, by states*. Washington, DC: National Academies Press. *www.nextgenscience.org/next-generation-science-standards.*

Virginia Department of Education, Office of Humanities and Early Childhood. 2013. *Virginia's foundation blocks for early learning: Comprehensive standards for four-year-olds. www.doe.virginia.gov/instruction/early_childhood/preschool_initiative/foundationblocks.pdf.*

Chapter 9

Water Play

Water play is so common in early childhood programs that it deserves its own chapter, although the concepts involved are also part of activities in other topic sections (see "Objects in Motion," "What Are Things Made Of?" "Mixing and Making a Change," and "Weather" in Table 1.1, pp. 15–39). A beginning understanding of forces (pushes and pulls), the properties of liquids, and the interaction between water and air (bubbles) develops during explorations involving water. Although understanding the molecular basis for water's unique chemical and physical properties is beyond young children, hands-on explorations involving water introduce them to the properties of liquid and solid water such as adhesion and cohesion. Plants and animals need water to survive, water can change the shape of the Earth, water is present in weather events—these are all science ideas that young children can understand. Investigations about water can expand in so many directions that I urge you to use a resource such as *Exploring Water With Young Children* (part of the Young Scientist Series; (Chalufour and Worth 2005) that is dedicated to this topic.

Water play may happen wherever children and water come in contact—at the bathroom sink, in a rain shower, at snack time, or while watering plants. As children engage in open exploration at a water table, or in a puddle in the outdoor play area, they may notice interesting phenomena such as drops sticking to their hands or tools, or the way water beads up on some but not all surfaces. Teachers support developing science inquiry by using open-ended questioning, providing additional materials at crucial times, promoting documentation of observations, and encouraging discussion that reflects on ideas. These strategies help deepen children's understanding of the properties of water and the processes that affect it.

An inquiry into the properties of water can include exploring the flow of water down and up through tubes (liquids take the shape of the container), building with frozen water (water can be solid if cold enough), and investigating the forms of water as precipitation (water falls from clouds), among other activities. Build on children's open exploration by observing them closely so you can be ready to share information from trade books or give them an interesting new tool, support that will extend their investigation into science concepts about water.

Children pay close attention to the properties of water when they work with small amounts using small tools such as spoons and droppers (see Chapters 24, "Water Works," p. 141, and 41, "Ongoing Inquiry," p. 243). After observing water drops sticking together and to some surfaces, children can explain why wet sand is a better building material than dry sand (Chapter 37, "Building With Sand," p. 220). The force of water's push can be explored with sand (Chapter 52, "Water Leaves 'Footprints,'" p. 311), in blowing bubbles (Chapter 39, "Developing Observation Skills," p. 232), and as a force to channel (Chapter 36, "Helper Hats," p. 214). The amazing material called water is everywhere—it can be liquid or solid (Chapter 15, "The Matter of Melting," p. 88), is essential for living organisms (Chapter 40, "Investigable Questions," p. 237), and falls from the sky (Chapter 33, "Adding Up the Rain," p. 197). As children explore the properties of water, they build a foundational knowledge of a material that is important in all the areas of science. Additional questions about water as weather phenomena are addressed in Chapter 10, "Weather."

Reference

Chalufour, I., and K. Worth. 2005. *Exploring water with young children.* Young Scientist Series. St. Paul, MN: Redleaf Press.

Chapter 10

Weather

Weather events such as changing temperature, moving clouds, wind, and precipitation are central to the lives ·of young children because they happen every day and are easily observable. Some are notable because of the changes to the daily routine—precipitation or relatively extreme temperatures may make adults reluctant to allow outdoor play, and severe events may endanger the community. Daily weather events reflect changes in the system of the Earth's atmosphere. Learning about the weather is part of Earth science, a science that also includes investigating water, soil, and rocks, as well as the movement of the Earth in space as observed in the patterns of movement of the Sun, Moon, and stars (Massachusetts Executive Office of Education [Massachusetts EOE] 2013; NGSS Lead States 2013).

Young children develop beginning ideas or naïve theories about weather events based on their experiences. They may not be sure where ice comes from and may think moving tree branches cause the wind. "Mistaken ideas may be the only plausible way for a child to progress toward a more accurate understanding of scientific concepts" (Michaels, Shouse, and Schweingruber 2008, p. 44). Regular observation and recording of temperature (relative or measured), cloud cover, precipitation, and wind strength focus children's attention on weather over time and allow them to find patterns such as "Which season has the coldest temperatures?" "Are there always clouds present when it's raining?" and "Does wind blow constantly?"

Your area may have more of one kind of weather event than of others. Providing a structure for finding a pattern in weather observations, such as comparing the children's weather data between one month and the next, guides children's understanding of the disciplinary core ideas in Earth science and the crosscutting concepts of identifying patterns and recognizing systems (NGSS Lead States 2013). A graph of clothing observations reveals changes in temperature as children add or remove layers in response to seasonal weather changes (Ashbrook 2015), and a weekly count of sunny and rainy days over the year tells when the rainy season begins. The following are some of the core science ideas children can investigate about weather events:

- Weather is the combination of sunlight, wind, snow or rain, and temperature in a particular region and time. People record weather patterns over time (NGSS Lead States 2013).

- Simple tools can be used to collect and record data on elements of daily weather, including sun or clouds, wind, snow or rain, and higher or lower temperature; describe how local weather changes from day to day and over the seasons; and recognize patterns in those changes (Massachusetts EOE).

- Weather has an impact on living things (Massachusetts EOE).

The idea of temperature is introduced when you begin with an activity that will establish a shared vocabulary for effects of temperature changes (see Chapter 15, "The Matter of Melting," p. 88). Chapter 43, "Measuring Learning" (p. 255), presents a way to relatively measure the temperatures of outdoor spaces and playground structures. Children are often familiar with the concept of using numbers to describe a temperature from having their temperature taken in case of fever. They can use technology—analog and digital thermometers—and watch the red line go up or down, or watch the numbers change as they move a thermometer from one location to another (Chapter 50, "The Wonders of Weather," p. 295). During the hours spent outside, brief conversations about the natural phenomena in the surroundings introduce key aspects of weather for further exploration—the Sun and air (Chapter 22, "The Sun's Energy," p. 130, and Chapter 29, "Air Is Not Nothing," p. 171). Playing in precipitation of any kind connects children's experiences with the properties of water (Chapter 41, "Ongoing Inquiry," p. 243) and prompts interest in "how much rain?" (Chapter 33, "Adding Up the Rain," p. 196).

Talking about the weather is a daily opportunity to increase children's awareness of the natural world and the way they can participate in documenting it through collecting data. Over time, these moments will allow children to see patterns in their data—when was the biggest snowfall, the coldest week, the month without rain, and the season that is best for growing watermelon?

References

Ashbrook, P. 2015. About the weather. *Science and Children* 53 (1): 4–5.

Massachusetts Executive Office of Education (Massachusetts EOE). 2013. *Pre-K science, technology and engineering standards. www.mass.gov/edu/birth-grade-12/early-education-and-care/curriculum-and-learning.*

Michaels, S., A. W. Shouse, and H. Schweingruber. 2008. *Ready, set, SCIENCE! Putting research to work in K–8 science classrooms.* Washington, DC: National Academies Press.

NGSS Lead States. 2013. Appendix E—Progressions within the *Next Generation Science Standards*. In *Next Generation Science Standards: For states, by states.* Washington, DC: National Academies Press. *www.nextgenscience.org/next-generation-science-standards.*

Additional Resource

Henriques, L. 2000. Children's misconceptions about weather: A review of the literature. Paper presented at the annual meeting of the National Association of Research in Science Teaching, New Orleans, LA, April 29. *www.csulb.edu/~lhenriqu/NARST2000.htm.*

Miller, H., M. M. Smith, and K. C. Trundle. 2014. What's the weather like today? *Science and Children* 51 (5): 50–54.

Chapter 11

Designing and Building to Solve a Problem: Engineering

Although the acronym "STEM" (science, technology, engineering, and mathematics) was first used in the 1990s and has been commonly used in recent years, the engineering part of STEM is often neglected in early childhood programs. Educators are seeking to change this by focusing increased attention on developing integrated STEM teaching. Some aspects, or practices, of engineering (NGSS Lead States 2013) may already be part of your teaching practice:

- Helping children state the problem to be solved ("There aren't any blocks left and I need some to build my house")

- Recognizing the limits or constraints on the problem ("We only have one set of that kind of block")

- Representing a design with a drawing to help you plan and share your idea with other people ("Can you draw a picture of what you want your house to look like?")

- Supporting students to try different solutions to problems and evaluate them ("Are there other materials you could try using to see if they would work?")

- Helping children understand that failures are useful because they can be analyzed to see how to improve the design ("I see that the cardboard pieces you used for a roof fell off. What could you do to keep them attached?")

Children engage in beginning engineering design and problem solving as they play. They design a tool when they repurpose a pie pan as a paddle to bat pom-poms into the air in a game of catch. They try more than one way to solve a problem when they first use a toy train track, and then a long block, to pull toys out from under a bookcase. Teachers can add direction to these unstructured efforts without taking over from the children. The goal is to expand the learning without stopping the children's initiative. Teachers find that students are more motivated and engaged when they are working on a problem that relates to their world (Capobianco, Nyquist, and Tyrie 2013). Preparing the environment, making time for your own science and engineering investigations, and using a framework to ensure that engineering explorations include testing and redesign, as well as integrating literacy and mathematics skills, are some recommended strategies for teaching engineering (Hoisington and Winokur 2015). To learn more about the profession of engineering and how to teach the practices of engineering in your program, read children's books and consult teacher resources (see Appendix 1, pp. 345–346).

Children enjoy playing with water, so giving them an engineering challenge that requires water—for example, designing a boat that will float and carry a set of coins—will hold their interest through a period of design, test, and redesign. As they work with common materials such as paper, tape, corks, craft sticks, aluminum foil, sticks, and repurposed packaging materials, they will begin to understand how all the properties of the materials (not only shape) affect the success or failure. It is important for children to experience an engineering design process where they can learn from failures and build on their past successes (Tank et al. 2013).

You can model the acceptance of design failure as a way to find out how to improve your design any time you build with blocks and create an unstable structure: "My building keeps falling. I have to figure out what needs to be changed in my design." You can also call attention to the idea of constraints, limits on the properties of the materials, and the amounts of materials, space, or other resources available to implement your design (see Chapter 53, "The STEM of Inquiry," p. 316). Open-ended use of materials (Chapter 54, "Are They Getting It?" p. 322) and teacher-given problems (Chapters 17, "Roll With It," p. 99; 24, "Water Works," p. 141; and 27, "Objects in Motion," p. 160) connect children with the science concepts of motion, form and function, and properties of matter while using an engineering design process (Chapters 36, "Helper Hats," p. 214; 37, "Building With Sand," p. 220; and 48, "Drawing Movement," p. 282). Discussions about design and engineering process questions are too exciting and central to practicing problem solving to be confined to a unit. Instead, these reflections on the work (play) and questions about how to solve a problem should be a daily occurrence wherever children learn.

References

Capobianco, B. M., C. Nyquist, and N. Tyrie. 2013. Shedding light on engineering design. *Science and Children* 50 (5): 58–64.

Hoisington, C, and J. Winokur. 2015. Gimme an "E"! Seven strategies for supporting the "E" in young children's STEM learning. *Science and Children* 54 (1): 3–10.

NGSS Lead States. 2013. *Next Generation Science Standards: For states, by states.* Washington, DC: National Academies Press. *www.nextgenscience.org/next-generation-science-standards.*

Tank, K., C. Pettis, T. Moore, and A. Fehr. 2013. Hamsters, picture books, and engineering design. *Science and Children* 50 (9): 59–63.

Chapter 12

Technology

A *Framework for K–12 Science Education* defines technology this way: "Technologies result when engineers apply their understanding of the natural world and of human behavior to design ways to satisfy human needs and wants. This is not to say that science necessarily precedes technology; throughout history, advances in scientific understanding often have been driven by engineers' questions as they work to design new or improved machines or systems" (NRC 2012, p. 12). Children are compelled to solve problems in their play, sometimes using tools.

Science and engineering practices (NGSS Lead States 2013) include the use of tools to observe, measure, record, solve a problem, serve as a model, and build. Using tools helps people extend their senses to make observations that are more detailed or more exact than their senses are able to make. Using tools and thinking about their form provide experiences for reflecting on the design of tools. As children become familiar with tools and their shape, they begin to understand the limits of their senses and the properties of materials used in the tools—for example, the lenses of magnifiers are curved, units of measurement are uniform, wheels are round, and techniques for measuring and applying heat are useful in science explorations. People also use tools, such as pencil and paper or photography, to document their observations.

Electronic technology is beyond the scope of *The Early Years* column, but I encourage you to explore the ways that cameras, digital microscopes, sensors, tablets, and phones can add to your students' understanding of using tools to investigate the world and document their learning (see the "Additional Resources" section at the end of this chapter). As digital media become more easily available and present in early childhood programs, the American Academy of Pediatrics urges educators to "guide students to engage in appropriate, positive, and safe ways to utilize these helpful digital resources" (American Academy of Pediatrics 2015).

When tools such as magnifiers, flashlights, lengths of links, or "measuring hands" (Ashbrook 2006) and containers are readily available to children, they become used to getting the tools when they need them rather than waiting for an adult to suggest it. Even 3-year-olds will begin to call for a container and magnifier when they see an insect. Children may explore tools in ways that seem unproductive to adults, such as holding a magnifier very close to the eye. This can continue as long as they don't damage their eyes or the tools or reject the usefulness of magnifiers. Children's exploration over time teaches them how to position the magnifier for best viewing. Offer suggestions about where to hold the magnifier and where to position the head if they become frustrated or seem puzzled.

When first exploring magnification, children may think the object actually gets bigger rather than just looking bigger. With a productive question from a teacher ("What does the object look like when you look around the magnifier?"), discussion (not lecture), multiple tries, and viewing from different angles, children come to understand that the object's actual size does not change.

Magnifiers are a commonly recognized science tool. Review Chapter 26, "Observing With Magnifiers" (p. 154) and the other chapters listed in the "Technology" topic section of Table 1.1 (pp. 15–39) to see how technology can be part of the explorations and investigations that interest your students. Models and maps (Chapter 44, "A Sense of Place," p. 261), tools (Chapter 24, "Water Works," p. 141, and Chapter 50, "The Wonders of Weather," p. 295), and materials used in solving problems (Chapter 17, "Roll With It," p. 99) are ways for children to begin using technology.

References

American Academy of Pediatrics. 2015. Growing up digital: Media research symposium. *www.aap.org/en-us/Documents/digital_media_symposium_proceedings.pdf.*

Ashbrook, P. 2006. Learning measurement. *Science and Children* 44 (2): 44–46.

National Research Council. 2012. *A framework for K–12 science education: Practices, crosscutting concepts, and core ideas.* Washington, DC: National Academies Press. *www.nap.edu/catalog/13165/a-framework-for-k-12-science-education-practices-crosscutting-concepts.*

NGSS Lead States. 2013. Appendix F—Science and engineering practices in the *Next Generation Science Standards.* In *Next Generation Science Standards: For states, by states.* Washington, DC: National Academies Press. *www.nextgenscience.org/next-generation-science-standards.*

Additional Resources

Brown, A., D. L. Shifrin, and D. L. Hill; American Academy of Pediatrics. 2015. Beyond "turn it off": How to advise families on media use. *AAP News*, September 8. *www. aappublications.org/content/36/10/54.full*.

Blagojevic, B., and K. Thomes. 2008. Young photographers: Can 4-year-olds use a digital camera as a tool for learning? *Young Children* 63 (5): 66–72.

Coskie, T. L., and K. J. Davis. 2009. Larger than life: Introducing magnifiers. 2009. *Science and Children* 46 (9): 64–66.

National Association for the Education of Young Children. 2012. Resources for technology and young children: New tools and strategies for teachers and learners. *Young Children* 67 (3): 12–13, 58.

National Association for the Education of Young Children (NAEYC). 2015. Technology and young children: NAEYC interest forum. *www.techandyoungchildren.org*.

National Association for the Education of Young Children (NAEYC), Technology and Young Children Interest Forum. 2008. On our minds: Meaningful technology integration in early learning environments. *Young Children* 63 (5): 48–50.

Parnell, W., and J. Bartlett. 2012. iDocument: How smartphones and tablets are changing documentation in preschool and primary classrooms. *Young Children* 67 (3): 50–57.

Sneider, C. 2012. Core ideas of engineering and technology: Understanding *A Framework for K–12 Science Education. Science and Children* 49 (5): 8–12.

Part III.

The Early Years Column and Companion Blog

Chapter 13

Introduction to *The Early Years* Column and Companion Blog

Early childhood science teaching may begin with a child's question, a teacher's desire to share the excitement of discovery, or a program curriculum with a series of activities. Children's interests often branch out from the initial exploration, and teachers may have to research content information, teaching strategies, and standards to continue providing engaging conversation and activities to explore the concepts involved.

The concepts covered in *The Early Years* column, which is published in *Science and Children*, are further explored in a companion blog called *Early Years*. The blog posts provide additional stories, discussion, and resources to support your teaching about the relevant topic or concept. Comments and questions can be posted at *www.nsta.org/earlyyears*.

Chapters 14–57 of this book contain the selected column entries in chronological order by original date of publication, as well as edited excerpts from and links to related *Early Years* blog posts. The column entries have been edited and updated to include the *Next Generation Science Standards (NGSS)*, safety information, and additional resources. The blog posts can be accessed through the addresses in each chapter and are also listed at *www.nsta.org/earlyyearsbook*. (The quick response [QR] code to the right will take you to the list.) The column and the blog are also archived online by the National Science Teachers Association; see *http://learningcenter.nsta.org/share.aspx?id=kWe5LKQ7hf* for a collection of column entries and *http://nstacommunities.org/blog/category/earlyyears* for the blog archive.

Chapter 14

What Can Young Children Do as Scientists?[1]

BEFORE YOU BEGIN THIS ACTIVITY:

Remember that each activity is not a stand-alone science or engineering curriculum. Activities are small steps in a journey of science inquiry, as discussed in Chapter 1. Your students will learn more about this concept and about the nature of science if you use this activity as part of an ongoing exploration of a question, a concept, or a topic being investigated by your class. Ask yourself, "What should come before this and what should come after?" Refer to Table 1.1 (pp. 15–39) to find other activities from The Early Years *column that address the same concepts.*

Children come to our classrooms with varying amounts of experience in exploring the natural world and participating in scientific inquiry. The school year's first meeting for science instruction can be a time to take stock of the breadth and depth of your students' experiences with science.

Learning about the "nature of science" is part of the *Next Generation Science Standards* (NGSS Lead States 2013), which state that "Science is a human endeavor. ... Men and women of diverse backgrounds are scientists and engineers. Reading the book *What Is a Scientist?* by Barbara Lehn (1999) is an engaging way to help students understand your expectations of them as scientists and what to expect when it's "Science Time" in the classroom. It's also a gentle reminder to teachers that, for the most part, children can participate fully in the process of inquiry at age-appropriate levels. Simplifying the process, or even the vocabulary, is a

---

[1] This column entry was originally published in *Science and Children* in September 2005.

disservice to the students' interest and ability to observe closely, ask questions, wonder, use tools, collect data, use logical thinking, consider alternative explanations, record findings, share information, and build on new experiences to develop new ideas about the world.

The book begins with the statement, "A scientist is a person who asks questions and tries different ways to answer them." The photographs on the facing pages are accompanied by descriptions of the pictured inquiry, such as, "Hannah wonders about fruit and vegetables. 'Are there this many peas in every pea pod?' asks Hannah." The two levels of text make the book appropriate to use across many grade levels. The book concludes with a statement we can all relate to, "A scientist has fun."

One way to use this book is to post pages with each statement from the book around the classroom. As they conduct various science activities in class, children can record the activity on the appropriate page with photographs, drawings, or descriptive paragraphs that they have written or dictated.

References

Lehn, B. (with photographs by Carol Krauss). 1999. *What is a scientist?* Brookfield, CT: Millbrook Press.

NGSS Lead States. 2013. Appendix H—Understanding the scientific enterprise: The nature of science in the *Next Generation Science Standards. Next Generation Science Standards: For states, by states.* Washington, DC: National Academies Press. *www.nextgenscience.org/next-generation-science-standards.*

Activity: Scientists at Work

Objective
To relate students' work to the work of "real scientists" by looking at many photographs of scientists in action

Materials
* *What Is a Scientist?* by Barbara Lehn

- Photographs of people engaged in scientific work: measuring, looking, writing, using tools, exploring, building, comparing, and talking. Try to find at least one photograph for each statement in *What Is a Scientist?* Magazines are the best source and can be supplemented with historical photos from internet sites. With careful attention, your collection can include a rainbow of peoples of all ages and many cultures. Lamination will enable the photos to survive frequent handling.

Procedure

1. Hand your students many photographs of people engaged in scientific activities.

2. Ask open-ended questions to get the children talking: *What are these people doing? Why are they doing that?* Let the children's answers stand without correction. If they ask, give further general information: *That scientist is measuring. Those scientists might be digging to find fossils to help us understand what Earth was like before there were people. She is trying to find out something about that plant she is looking at, so she can make a medicine from it.*

3. Invite the students to reminisce about science activities they have been involved with by asking: *Have you done anything like that?*

4. Talk about what students might do when they grow up: *Maybe when you grow up you will do such activities as* (complete this statement with whatever is suggested by their descriptions and the photographs, e.g., *write about insects, help grow new plants, find out how to keep people healthy, go to space to learn how to live there, build with materials you invent, study how people lived long ago*).

Children will begin to relate their own science activities and experiments to the pursuit of scientific knowledge by adults. Photographs of doctors examining people and of scientists in the field wearing work clothes appropriate for their specialized activities may surprise students who have become accustomed to scientists being represented by the image of a white man wearing a white lab coat in a laboratory. Through discussion about the photographs, a teacher can broaden the children's ideas of what constitutes science and what a scientist looks like.

Extension

Use one of the following sorting activities to learn what students understand about the process of scientific inquiry:

1. Have the students each choose one photograph and match it with one of the activities listed in *What Is a Scientist?* Ask the students to share why they grouped it that way.

2. Have the students sort the photographs into other groups of their own choosing to learn what the students think science is all about.

Resources

Lehn, B. (with photographs by Carol Krauss). 1999. *What Is a Scientist?* Brookfield, CT: Millbrook Press.

National Association for the Education of Young Children (NAEYC) and the National Association of Early Childhood Specialists in State Departments of Education (NAECS/SDE). 2002. *Early learning standards: Creating the conditions for success.* Washington, DC: NAEYC and NAECS/SDE.

National Research Council. 1996. *National science education standards.* Washington, DC: National Academies Press.

Teacher's Picks[2]

Publications

Everyone Is a Scientist by Lisa Trumbauer (Capstone Press, 2001).

Written for emergent and early-level readers with photographs that show an adult scientist at work and a student scientist at work, this book is a good lead-in to many activities:

- Make a list of what things a scientist could study. What kind of tools do scientists use?

- Make a list of what a scientist could measure. Ask students to choose something they are interested in and write science facts in their journal or draw a picture.

- Take photographs of students using a tool, looking at things closely, measuring, writing down facts, and asking questions. Use the photographs to make a class book titled "Everyone Is a Scientist" to share with their families.

---

[2] These suggestions were provided by Charlene K. Dindo, a marine and environmental science enrichment teacher.

Dancing on the Sand: A Story of an Atlantic Blue Crab by Kathleen M. Hollenbeck (Soundprints, 1999).

The life cycle of the blue crab is explained with detailed illustrations. A companion cassette recording with sound effects is great for capturing the attention of young scientists!

The Seaside Naturalist by Deborah Coulombe (Fireside, 1992).

This book has great black-line drawings and easy-to-understand text on sea creatures. Field guides are valuable for nonreaders or emerging readers because they can match shapes, look for similar colors and patterns, and compare animals. Using field guides, the internet, and books helps to validate the learning experience and confers ownership in what young children "discover!"

Websites

Seascape Video
www.seascapevideo.com/stock_footage/VideoClips.html

These 15- to 30-second clips demonstrate movement, camouflage, and color patterns of sea creatures such as flounder, sea turtle, and sea fans. Easy repeated viewing on the internet reinforces perception skills and offers a real-life connection to the topic or animal.

NOAA Photo Library
www.photolib.noaa.gov/nmfs

This website of the National Oceanic and Atmospheric Administration (NOAA) has atlases of illustrations from a study done in the 1880s by the U.S. Commission of Fish and Fisheries. Download an exquisite scientific illustration of the blue crab (and many other sea animals) from the "Natural History of Useful Aquatic Animals" on this website. Students will easily notice similarities and differences between animals with the degree of detail in these images.

Related *Early Years* Blog Posts

The following information is from "Displaying Children's Science Learning," published September 11, 2008:

> *Children love to look at the photos and reminisce about past activities.*
> *Post your photos, or children's work, under headings borrowed from, or*

inspired by, Barbara Lehn's book What Is a Scientist? *Examples of these headings include the following:*

- *A scientist is a person who asks questions and tries different ways to answer them.*

- *A scientist learns from her senses.*

Read more at *http://nstacommunities.org/blog/2008/09/11/displaying-childrens-science-learning.*

Additional related blog posts:

"Celebrate Pink!" published June 23, 2009: *http://nstacommunities.org/blog/2009/06/23/celebrate-pink.*

"Early Childhood Science in Preschool—A Conversation on Lab Out Loud," published February 24, 2014: *http://nstacommunities.org/blog/2014/02/24/early-childhood-science-in-preschool-a-conversation-on-lab-out-loud.*

Chapter 15

The Matter of Melting[3]

BEFORE YOU BEGIN THIS ACTIVITY:

Remember that each activity is not a stand-alone science or engineering curriculum. Activities are small steps in a journey of science inquiry, as discussed in Chapter 1. Your students will learn more about this concept and about the nature of science if you use this activity as part of an ongoing exploration of a question, a concept, or a topic being investigated by your class. Ask yourself, "What should come before this and what should come after?" Refer to Table 1.1 (pp. 15–39) to find other activities from The Early Years column that address the same concepts.

The young child who asks, "Where is my ice cream?" when confronted by a bowl of milk an hour after leaving a bowl of ice cream may have a hard time grasping that a solid volcanic rock was once a red hot liquid. Unless they witness it, young children may not understand that solids can melt into liquids and liquids can become solids.

It is common for children to describe *dissolving* as *melting*. Perhaps the term *melting* is more familiar than *dissolving,* and children often use a word they know to mean a change took place. Doing melting and dissolving activities back-to-back would illustrate how the two processes are different. And keeping a child's attention while a solid melts completely can be a challenge. That is why when exploring melting, it's worth it to repeat the experience a few times with various substances, including chocolate and wax.

Bringing these experiences into the classroom makes them part of the classroom culture and provides an opportunity to teach vocabulary and begin building your shared language. Reading a story about one of these experiences will help students connect something from their own lives to the concept. "The warm wax is soft," and "My hand melted the chocolate" are statements about shared experiences that, when grouped together, present the concept of states of matter. The language established by a shared experience is a foundation from which to begin teaching science, building toward the *Next Generation Science Standards* grade 2 performance expectation 2-PS-1 Matter and Its Interactions (NGSS Lead States 2013).

---

[3] This column entry was originally published in *Science and Children* in January 2006.

Investigations into melting involve more than learning about changes in state due to changes in temperature. Making predictions; using their senses (and tools such as a thermometer) to gather information; observing; describing and recording changes; and sharing explanations are science processes young students will gain practice in through this exploration.

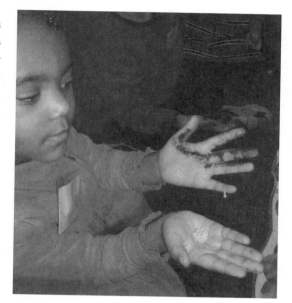

Reference

NGSS Lead States. 2013. *Next Generation Science Standards: For states, by states.* Washington, DC: National Academies Press. *www.nextgenscience.org/ next-generation-science-standards.*

Activity: Melt Away!

Objective

To learn that heating a solid can make a change in states called *melting,* which can be seen, felt, and measured

Materials

- Beeswax (available at craft stores)

- Chocolate chips

- Thermometers (optional)

CAUTION: Check with parents to see if students have allergies to chocolate or dairy products before allowing them to handle the chocolate chips. There are brands of chocolate chips that are made without dairy products. Remind students that in a science laboratory no eating is allowed, but they will be able to eat the chocolate chips after the investigation.

Procedure

1. To get students interested in the topic and introduce or review vocabulary, talk about the states of matter—liquid, solid, and gas—throughout the day. For example, after taking a deep breath you can say, *Oh, it feels good to breathe in this air. I'm glad it's a gas, not a liquid!* When cleaning up after an activity, you can say, *I'm washing my hands with liquid water.* Or you can take

advantage of a minor mishap by saying *Ouch! The block that dropped on my foot is a hard solid.*

2. Give each child a small piece of beeswax to feel and hold. Children often roll it into a ball. Have them hold it in one hand and give them each a chocolate chip to hold in the other hand. Put one chip and a piece of wax on a table in the children's view.

3. Have the students describe the two substances, including which state of matter they fall into (liquid or solid), and record their words. Ask students to predict what will happen to the chocolate chip and the beeswax piece in their closed hands and to the same items on the table. Record the students' predictions.

4. Jump in place or march around the room so that at least some of the children's hands will become warm enough to melt the chocolate chip. Ask a question: *Will there be any changes to these materials?* After 2–5 minutes, have all the children open their hands. To reassure children who are concerned about messy hands, express positive excitement about any changes. Some chips will be more melted than others. Ask the students to describe what the chocolate chip and ball of wax feel like now.

5. Ask questions such as, *Did any change happen to the wax or chocolate?* and *Which is softer?* to begin a group discussion about any changes. This will lead to a discussion about melting. Have the children define the term *melting* and list other substances that they have seen change from a solid into a liquid, such as ice, ice cream, snow, or candles.

6. If they are familiar with thermometers, have the students measure and compare the closed-hand temperature of children whose chips melted a lot with the closed-hand temperature of children whose chips did not melt.

Extension
Extend the exploration of changes in states of matter with these activities and discussion about children's observations:

*   Observe the change in state in the other direction (i.e., from liquid to solid) by making juice pops from frozen juice concentrate. Have the students feel the concentrate before thawing and then observe the solid after you make the pops.

*   Freeze liquid paint in ice-cube trays, adding craft sticks for handles. Paint with the "melting" cubes.

- Melt a bar of glycerin soap in a clear glass bowl in a microwave oven. Have students watch the bar as it changes shape. Cool the liquid soap in an empty plastic margarine tub. *CAUTION: Hot liquid soap should not be carried around the classroom.*

Observing the melting of several different substances will strengthen students' understanding that melting is a change in state from solid to liquid that can happen to many substances when heated, not a characteristic solely of the familiar ice. It's fun to see that amazing change, and it is important that young children get a range of experience with melting before learning in later grades about molecular structure differences between states of matter.

Teacher's Picks

Look for fiction as well as nonfiction resources to extend students' thinking about scientific concepts.

Publications

Ice Palace by Deborah Blumenthal (Clarion, 2003).

This story of an annual winter carnival shows how ice can be used as a building material.

What Is the World Made Of? All About Solids, Liquids, and Gases by Kathleen Weidner Zoehfeld, illustrated by Paul Meisel (Let's-Read-and-Find-Out Science, Stage 2; HarperCollins, 1998).

Take students on a "field trip" with a class that—just like your class—is engaged in finding out about solids, liquids, and gases.

Investigating Solids, Liquids, and Gases With Toys: States of Matter and Changes of State by Lynn Hogue, Mickey Sarquis, Linda Woodward, and Jerry L. Sarquis (McGraw-Hill, 1997).

Use this book of middle-level activities as a resource to keep just ahead of your students' knowledge.

The Snowy Day by Ezra Jack Keats (Viking Juvenile, 1962).

This classic story recounting the many joys of playing in snow includes the classic childhood move of putting a snowball in a pocket to save for later and then discovering only a wet spot when looking for it later.

Websites

Watch a Glacier Melt

www.amnh.org/explore/science-bulletins/earth/documentaries/archived-in-ice-rescuing-the-climate-record/interactive-watch-a-glacier-melt

This interactive resource on the American Museum of Natural History website depicts melting as seen in nature.

Changing State

www.bbc.co.uk/schools/scienceclips/ages/9_10/changing_state.shtml

Designed for 9- and 10-year-olds, this animation of changes in state on the BBC website can be used as a review of the actual experience.

Related *Early Years* Blog Posts

The following information is from "Exploring the Properties of Liquid, and Solid, Water," published May 15, 2014:

> *Young children experience a sense of wonder when handling a block of ice that has an object embedded in it. How did the object get into the ice? They may be very familiar with the frozen water as ice cubes, on icy sidewalks, or as icicles, but they don't really know the conditions in which water freezes to ice. Experiences observing changes in water as the temperature changes build children's understanding of the properties of water and the reversible changes of freezing and melting.*

Read more at *http://nstacommunities.org/blog/2014/05/15/exploring-the-properties-of-liquid-and-solid-water.*

Additional related blog posts:

"Sensory Table Explorations of Matter," published December 19, 2013: *http://nstacommunities.org/blog/2013/12/19/sensory-table-explorations-of-matter.*

"Snow Explorations," published December 26, 2009: *http://nstacommunities.org/blog/2009/12/26/snow-explorations.*

Chapter 16

Feet First[4]

BEFORE YOU BEGIN THIS ACTIVITY:

Remember that each activity is not a stand-alone science or engineering curriculum. Activities are small steps in a journey of science inquiry, as discussed in Chapter 1. Your students will learn more about this concept and about the nature of science if you use this activity as part of an ongoing exploration of a question, a concept, or a topic being investigated by your class. Ask yourself, "What should come before this and what should come after?" Refer to Table 1.1 (pp. 15–39) to find other activities from The Early Years column that address the same concepts.

Children love animals—and learning about them. Learning how animals use their external parts to help them survive is part of a progression that builds understanding toward the *Next Generation Science Standards* (*NGSS*) grade 1 performance expectation 1-LS1-1: "Use materials to design a solution to a human problem by mimicking how plants and/or animals use their external parts to help them survive, grow, and meet their needs" (NGSS Lead States 2013). Understanding how the shape of an object helps it function is part of the *NGSS* engineering performance expectation K-2-ETS1-2. Investigating the differences among various species of an "animal part" such as

---

[4] This column entry was originally published in *Science and Children* in April 2006.

feet can also be an intriguing way to help students develop classification skills and begin to make connections about the relationships between structure and function.

Although backbones are more pertinent to scientific classification, feet are more accessible. Children have direct experience with their own feet—what child hasn't experimented with making footprints in sand, mud, or snow? And, young children love an excuse to examine their own body and compare it with others.

Begin by discussing "What is a foot?"—cover the various meanings of the word (e.g., "the foot of the stairs," "the feet of a couch," and "4 feet tall") to make sure everyone is thinking of the same kind of feet. Because children often describe something by its function, the discussion will naturally move to discussion of "What are animal feet used for?"

Next, as a class, compile a list of how feet are used. Students will suggest walking, running, climbing, kicking, scratching, and digging, among other uses. Responses such as "Squirrels run, jump, and climb" and "Horses walk, run, and jump, but not climb" can be used to point out differences in foot structure as students think about why squirrels can climb. After listing all the uses of feet, bring out models of animals (toys) and animal track identification books (e.g., Murie and Elbroch 2005) to help the children investigate and describe the specifics of a particular animal's feet. Use realistic models of a variety of animals, including humans and other mammals, birds, reptiles, and invertebrates such as insects and snails, to really get the children thinking about locomotion. Pictures can be used, but models are best as long as they are accurate.

When looking at a model, students can see if the animal has any feet and how many, if there are toes, how long the toes are, and if the front and hind feet are similar. They can begin comparing animals and voice their ideas about how a particular kind of body structure allows an animal to have a particular kind of movement. "A squirrel can climb because it has toes with claws that can grab onto a tree, but a horse's toes can't grab." "Snakes don't have feet but they can move." Comparing the feet to the tracks in the identification book shows how each type of foot makes a print—what part of the foot or body touches the ground.

Introduce the terms *toe, claw, heel pad, hoof, cloven hoof*, and *dewclaw* as needed. By learning the terminology for parts of feet, children will both feel like scientists and learn how biologists use differences in *morphology*, the form and structure of an organism or one of its body parts, to classify animals. For example, there are many small animals that are called "bugs," but only those possessing six legs are in the class Insecta. In another example, deer and pigs (animals that walk on two toes and thus appear to have "cloven hooves") are in a separate order from horses and zebras (animals whose hooves are not "split").

Classification systems change as they incorporate new information learned by studying animal DNA. After examining the models and identification books, try the activities described later in this chapter to continue the foot exploration.

References

Murie, O. J., and M. Elbroch. 2005. *Animal tracks.* 3rd ed. Peterson Field Guides. New York: Houghton Mifflin.

NGSS Lead States. 2013. *Next Generation Science Standards: For states, by states.* Washington, DC: National Academies Press. *www.nextgenscience.org/next-generation-science-standards.*

Activity: Footprint Fun

Objective

To explore how feet differ from animal to animal and to think about how foot structure is related to animal behavior and habitat. Footprint Fun includes two activities, My Foot and Animal Print Stamps; the procedure for each is described after the list of materials.

Materials

- Playdough

- Paint in a shallow tray

- Small paintbrushes

- Paper towels

- Paper

- Animal track identification books (see "Resources" section) or internet information sheets

- Art foam (available from arts and crafts stores)

- Scissors

- Styrofoam

- White school glue

Procedure for My Foot

1. Have students remove one shoe and sock and look at their foot and describe it, either verbally or in writing. Prompt students with questions such as, *How many toes do you have? What can you do with your toes? Can you hold a crayon or scratch an itch with your toes? What part of your foot touches the floor when you walk, tiptoe, or jump?*

2. This part of the activity is easier to do in small groups. Spread newspapers on the floor and have a stack of paper towels nearby for cleanup so children can wipe the paint off their feet. Have some students, one at a time, put one foot into a thin layer of paint in a shallow tray and then step onto a piece of paper. Have them repeat the printmaking on tiptoe. Invite other students to step into playdough and make imprints of their foot. Have them repeat the imprint making on tiptoe.

3. Ask all the students, *Do the footprints look like you thought they would? Do you think the footprints look like those of any other animal?* Label the prints and imprints with "toe" and "heel" and save to use in the following activity.

Procedure for Animal Print Stamps

1. Using an animal track identification book (see "Resources" section), you can make animal footprint stamps and then use them for students to compare animal footprints, including their own. Choose animals with footprints large enough to see easily. Copy the footprints onto art foam (a thin, dense, but flexible foam that is easy to cut with scissors), cut out the shapes, glue the shapes to a Styrofoam-block base, and allow to dry overnight. Print each footprint stamp on a separate sheet of paper by brushing the art foam surface with paint and then pressing it to the paper. (Putting the stamp directly into the paint results in a sloppy print.)

2. Show the footprints to the students and ask them to describe the foot that made them. Prekindergarten students often respond at first with "Dinosaurs!" because they have heard of fossil dinosaur footprints. Challenge the students to identify the animals that made those types of footprints by matching them to those shown in an animal footprint identification book.

3. Raise awareness of how body structure relates to animal behavior and its environment by discussing questions over time: *Do all animals need feet? Why would an animal have a certain kind of foot? What are toenails useful for? How can we learn what different feet shapes are well suited for?* Children enjoy

National Science Teachers Association

counting and comparing the number of toes on footprints—a quality that is easy to define. They find it strange that some birds' toes point backward, leading to a discussion about how birds grip branches. Older students often comment on the prominent nails on the footprint of a big dog and how dogs dig in the dirt with paws, not a tool. The resemblance between a raccoon's front foot and a human hand leads children to speculate that raccoons can hold objects in much the same way humans can. After examining different bird footprints, students begin to associate webbed feet with birds that spend much of their time swimming in water.

For very young students, these activities may mean more about matching shapes and noticing differences. Older children, however, can begin to speculate about foot function and make predictions about habitat based on the shape of an animal's foot. Whatever the age of the student, comparing animal prints and reflecting on their own observations will deepen students' understanding of foot function *and* of science inquiry.

Resources

Arnosky, J. 2008. *Wild tracks! A guide to nature's footprints.* New York: Sterling Children's Books.

Dendy, L. 1996. *Tracks, scats and signs.* Take Along Guides. Minnetonka, MN: North Ward.

Murie, O. J., and M. Elbroch. 2005. *Animal tracks.* 3rd ed. Peterson Field Guides. New York: Houghton Mifflin.

Selsam, M., and M. H. Donnelly (illustrator). 1995. *Big tracks, little tracks: Following animal prints.* Let's-Read-and-Find-Out Science, Stage 1. New York: HarperCollins.

Teacher's Picks[5]

Publications

Forest Explorer: A Life-Size Field Guide by Nic Bishop (Scholastic, 2004).

This book works both as a guide to what to look for and a research tool to identify "what we saw." Composite photos showing life-size animals in forest scenes help train kids' eyes to spot real wildlife. A picture key identifies and gives natural history notes for all the animals in each double-page scene.

---

[5] These suggestions were provided by Sarah Glassco, a naturalist and early childhood science resource teacher.

One Small Place in a Tree by Barbara Brenner (HarperCollins, 2004).

Reading this history of a hole in a tree invites us to imagine all of the plants and animals that may have lived in and drawn life from the tree, even after it has died.

Pet Bugs: A Kid's Guide to Catching and Keeping Touchable Insects by Sally Kneidel (John Wiley & Sons, 1994) and *More Pet Bugs: A Kid's Guide to Catching and Keeping Insects and Other Small Creatures* by Sally Kneidel (John Wiley & Sons, 1999).

These readable "how-to" guides describe common insects and small creatures that are easy to catch and care for and safe to handle in the classroom.

Website

Animal Diversity Web
http://animaldiversity.org

This searchable encyclopedia sponsored by the University of Michigan Museum of Zoology is a good teacher resource for pictures and information on many species.

Related *Early Years* Blog Posts

The following information is from "Footprints in the Snow—Books to Extend Learning," published February 7, 2010:

> *Children enjoy finding footprints in snow, mud, or sand and guessing who made them. I made fake dinosaur footprints in the snow and the children noticed them (but nobody was fooled). Your children may enjoy making handprints and footprints in playdough indoors after exploring what prints they can find or make outdoors in snow, sand, or mud.*

Read more at *http://nstacommunities.org/blog/2010/02/07/impressions-in-the-snow.*

Additional related blog post:

"More Snow? Counting and Science in Winter Cold," published March 2, 2014: *http://nstacommunities.org/blog/2014/03/02/more-snow-counting-and-science-in-winter-cold.*

Chapter 17

Roll With It[6]

BEFORE YOU BEGIN THIS ACTIVITY:

Remember that each activity is not a stand-alone science or engineering curriculum. Activities are small steps in a journey of science inquiry, as discussed in Chapter 1. Your students will learn more about this concept and about the nature of science if you use this activity as part of an ongoing exploration of a question, a concept, or a topic being investigated by your class. Ask yourself, "What should come before this and what should come after?" Refer to Table 1.1 (pp. 15–39) to find other activities from *The Early Years* column that address the same concepts.

Sliding common objects, such as blocks, boxes, balls, empty containers, and even play foods, down a ramp is a fun way for children to start exploring physical science concepts related to the position and motion of objects and the crosscutting concept Structure and Function (NRC 2012). And, it's a great introduction to an exploration of the wheel, another familiar object that kids love and will be surprised to discover how it can be used to make work easier.

Let groups of students test a variety of small objects to see how they move. *Which objects can you blow across the table? How do they move— rolling or sliding? What shape moves (rolls) the easiest? Where do you see this shape in the world?* Students will discover that the small force of a breath of air will move only the lightest objects or those with a round cross-section.

Then, set up a simple ramp using a light-weight board or stiff cardboard with one end

---

[6] This column entry was originally published in *Science and Children* in July 2006.

set on top of several books and have students continue testing. Assign each group member a job, such as "Recorder" to record predictions and results; "Tester" to test if an object is a roller or a slider; "Chaser" to retrieve objects; and "Turn Keeper" to say which student goes next.

Before putting the object on the ramp, have the children predict which objects will move by rolling and which ones will move by sliding down the ramp. To record the predictions, the Recorder lists (and traces around) "Objects We Predict Will Roll" on one sheet of paper and "Objects We Predict Will Slide" on another sheet of paper (12 × 18 inches works well). Then the students can test their predictions by putting the objects, one at a time, at the top of the ramp and watching the result. The Recorder records the results by tracing the objects again on two new sheets of paper (one for "Rollers" and one for "Sliders"), so the group can compare the results with the predictions. Guide your students' recording if they are frustrated by the task. Having children participate in the data collection is more important than having a neat documentation sheet.

Afterward, discuss these questions with the class: *Did the object do what you predicted? What shapes are the objects that rolled? What shapes are the objects that slid?* Students will discover that a round, or nearly round, profile is necessary for rolling. To assess for understanding of how shape can predict movement, ask the students to draw a picture of one object in the room that would roll down the ramp and one that would slide down. Drawing to show how the shape of an object affects its function is part of the *Next Generation Science Standards* performance expectation K-2-ETS1-2 Engineering Design (NGSS Lead States 2013).

Next, take students to the development of the wheel by doing the following activity, which engages children in exploring the effects of pushes and pulls on the motion of an object, part of the *Next Generation Science Standards* kindergarten performance expectation K-PS2-1 Motion and Stability: Forces and Interactions (NGSS Lead States 2013). This activity is so much fun that all students will want to try it. Have them work in small groups of two to six. If materials are limited, have one group work at a time, out of sight of the others (if you have another adult to assist with supervision), to allow each group to make their own discoveries. The materials for exploring wheels take some time to prepare but can be used over and over again for many years.

References

National Research Council (NRC). 2012. *A framework for K–12 science education: Practices, crosscutting concepts, and core ideas.* Washington, DC: National Academies Press. *www. nap.edu/catalog/13165/a-framework-for-k-12-science-education-practices-crosscutting-concepts.*

NGSS Lead States. 2013. *Next Generation Science Standards: For states, by states.* Washington, DC: National Academies Press. *www.nextgenscience.org/next-generation-science-standards.*

Activity: Wheel Work

Objective

To experience the way work is made easier with the use of wheels

Materials

- One 50-lb bag of sand, tied inside an old pillowcase

- Two 25-lb bags of sand, each tied inside an old pillowcase

- A plywood board (about 2′ × 3′, sanded to reduce splintering) with a hole drilled through one end

- 10 feet of rope to loop through the hole in the board for the handle

- 8–14 tubes to use as "rollers," as wide as the board and between 8 cm and 12 cm in diameter (Flooring stores give away the cardboard tubes that carpet or linoleum come rolled on.)

- Wagon or tummy scooter used in physical education classes

Procedure

1. Tell students that you need help to move a heavy load (bags of sand) over a smooth surface (down the sidewalk or hall). As they volunteer, divide students into small groups of two to six students. Encourage the students to try moving the bags by pushing and dragging—first one at a time and then with the whole group working together.

2. Next, introduce the board as a tool to help move heavy loads. *CAUTION: Warn students to keep their hands away from the front of the board and out from under it when it is being pulled, to prevent pinched fingers.* Students will easily be able to push the bags onto the board and then drag the bags with the board by pulling the rope handle. Ask, *What takes less effort (is*

easier)—dragging the bag by itself or moving it by pulling it on the board? How much of the board is touching the ground when you pull the bags?

3.   Now bring the large "rollers" into view, and if no student suggests it, introduce the idea of using rollers under the board. *CAUTION: Warn students to keep their fingers out from under the board, to prevent pinched fingers.*

4.   Demonstrate how to lift one end of the board by the handle and slide tubes underneath the board to act as rollers. After a few trials, students will discover that the length of the tube needs to be positioned perpendicular to the board to work with the direction of pull on the handle. As the board rolls across the first few tubes, help students quickly put additional tubes in the path of the board to extend its "roll." As tubes are passed over, students move them to the front of the board to keep it rolling. Ask, *How much of the board is touching the ground, and how much of the rollers are touching the ground?* After all the groups have tried rolling the board over the tubes and have experienced how difficult it is for them, ask for ideas on how to make it even easier to move the bags.

5.   Some students may come up with the idea of a wheeled vehicle. If not, suggest the wagon or scooter. Ask, *Can this be moved by someone using just one finger?* The answer is usually yes! As each group moves the bags, repeat the question about the amount of contact between the wheels and the ground.

CAUTION: Do not let children ride on the board while another child pulls it, because children riding the board may easily fall backward and bump their head.

Discuss the various methods used to move the sandbags. *Which was easiest? Do you think the amount of touching between the bag or tool and the ground is related to how easy or hard it was to move the bags?* Have students draw pictures showing how much of the bag (and board, rollers, and wheels) touched the ground as they moved the bags. Comments such as, "I can move that bag all by myself, but wheels made it easy for me" and "We should have used the wagon first!" show that students know this simple machine is effective. Although they are not usually able to explain why until late elementary school, this activity gives the students an experience to build on.

Teacher's Picks[7]

Books for elementary physical science are hard to find, and those with accurate language and illustrations even harder; the ones listed here have been recognized as outstanding.

Publications

I Fall Down by Vicki Cobb (HarperCollins, 2004).

This visually appealing and conceptually sound physical science book introduces young children to the concepts of gravity and weight. The concepts are reinforced through many real-life examples and fun activities that could easily be duplicated in class or at home.

What Does a Wheel Do? by Jim Pipe (Copper Beech Books/Millbrook, 2002).

A series of questions about how and why things move are posed and then answered by simple investigations of shapes, surfaces, and slopes using ordinary materials. Investigations are extended in the "Solve the Puzzle" question that follows each explanation of "Why It Works."

Inclined Planes and Wedges; *Levers*; *Pulleys*; *Screws*; *Wheels and Axles*; and *Work* by Sally M. Walker and Roseann Feldmann (Early Bird Physics; Lerner, 2001).

These six books on simple machines not only have great photos of familiar objects to which children in grades K–4 can relate but also describe simple activities that can be replicated in the classroom.

Website

Inventor's Toolbox: The Elements of Machines
www.mos.org/sln/Leonardo/InventorsToolbox.html

This photo glossary of simple machines is part of a website devoted to Leonardo da Vinci that was created by the Museum of Science in Boston.

---

[7] These suggestions were provided by Juliana Texley, who has served in a variety of capacities for the National Science Teachers Association (NSTA), including chairing the committee that crafted NSTA's response to the 1996 National Science Education Standards; as editor of the NSTA journal *The Science Teacher*; as lead reviewer for NSTA Recommends; and as 2015–2016 president of NSTA.

Related *Early Years* Blog Posts

The following information is from "Spatial Thinking," published March 24, 2011:

> *To get the most out of this activity, children need time to investigate the relationship between the blocks, the slope of the ramps, and the size and weight of the balls. They need time to play ("Did the ball make it into the hole?"), time to compare ("The heavy ball bumped off the path here, but the light one kept going."), time to think and talk about why, and time to revise their structures ("I'm going to make it better this time!"). By designing, discussing their ideas, building, and revising their designs, I hope the children will gain experience with physical science concepts of force and motion while developing their spatial thinking.*

Read more at *http://nstacommunities.org/blog/2011/03/24/spatial-thinking*.

Additional related blog posts:

"Found Materials + Engineering Process = Toy," published April 26, 2012: *http://nstacommunities.org/blog/2012/04/26/found-materials-engineering-process-toy*.

"Preventing Misconceptions," published September 24, 2008: *http://nstacommunities.org/blog/2008/09/24/preventing-misconceptions*.

"Learning About Motion and Appropriate Restraints," published February 9, 2009: *http://nstacommunities.org/blog/2009/02/09/learning-about-motion-and-appropriate-restraints*.

"Are You Ready? (What I Learned on My Summer Vacation: Ramps, Video Conferencing With Children, and Climate)," published August 16, 2010: *http://nstacommunities.org/blog/2010/08/16/are-you-ready*.

Chapter 18

Young Questioners[8]

BEFORE YOU BEGIN THIS ACTIVITY:

Remember that each activity is not a stand-alone science or engineering curriculum. Activities are small steps in a journey of science inquiry, as discussed in Chapter 1. Your students will learn more about this concept and about the nature of science if you use this activity as part of an ongoing exploration of a question, a concept, or a topic being investigated by your class. Ask yourself, "What should come before this and what should come after?" Refer to Table 1.1 (pp. 15–39) to find other activities from The Early Years column that address the same concepts.

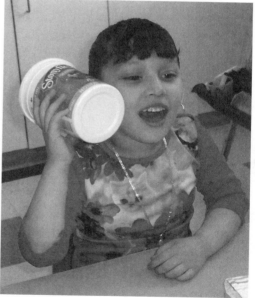

Children are often described as natural scientists, and their curiosity is seen as a basic human trait. They amaze adults with perceptive questions and extended explorations of the environment. Asking questions is one of the science and engineering practices identified in *A Framework for K–12 Science Education* (NRC 2012). Because asking questions is central to inquiry and learning in general, science teachers are especially invested in having all their students use this process skill.

However, some students do not ask questions—perhaps because of shyness, not realizing that they are allowed to, or lack of experience. Due to the normal range in development in young children and the differences in their early interactions with

---

[8] This column entry was originally published in *Science and Children* in September 2006. It was inspired by a segment of *3, 2, 1, Contact!* (a children's television science program), where the young hosts challenged Nobel Prize–winner chemist Linus Pauling to describe the object they had put in a small box. As he shook and turned the box, he listened to the sounds and told the children what he was thinking. His passion for discovering the unknown was obvious.

caregivers, the ability to frame a question is not a given in children entering kindergarten. Children may not answer questions for the same reasons.

There are many strategies to support and increase children's capacity for curiosity, but suggestions about teaching science tend to be vague, along the lines of "The teacher elicits more observations and questions. ..." How exactly can we develop and support questioning skills in students? One way to help all children learn to ask and answer questions is to invite the children to predict what they think will happen next in a book any time you read aloud. This type of practice is fun with the book *Fortunately* (Charlip 1993). The story has a pattern, and every page is an opportunity for students to ask (and answer) what will happen next in the book.

After reading a few pages of *Fortunately*, model asking a question: *I wonder if Ned is going to be able to get to the party?* or *I wonder if Ned is going to get hurt?* Make it clear that you want the children to ask questions and make predictions by saying so and making sure they all have enough time. In subsequent readings, a student can read the book (or tell the story) and invite the other students to ask, "What do you think will happen next?" and give answers.

Providing a space or time where children can ask questions supports students' questioning. Some structures that teachers use to show students that questions matter include a set quiet time for thinking, a question board (Pearce 1999), or a set of pages in a journal where children can post questions they have written down or dictated for future consideration. Working in small groups and pairing students are helpful for young children who lose their thoughts if required to wait very long to respond (Bailey and Brookes 2003). Showing respect for asking questions by listening and recording the questions encourages more questioning. Check your responses to children's questions for statements showing approval of asking questions; for example, you can say, *I like hearing your questions* and *You can really find out about things when you ask questions and try to answer them.*

With practice, children will make asking questions and trying to answer them a habit that carries over to a wide range of activities (e.g., "What's going to happen when I put this really big block on the top?" and "Why do the worms lie on the sidewalk when it rains?").

In the following activity, many questions may be asked and answered in small groups, one-on-one, or at a workstation for children to do the activity independently.

References

Bailey, B.A., and C. Brookes. 2003. Thinking out loud, development of private speech, and the implications for school success and self-control. *Young Children* 58 (5): 46–52.

Charlip, R. 1993. *Fortunately.* New York: Aladdin.

National Research Council (NRC). 2012. *A framework for K–12 science education: Practices, crosscutting concepts, and core ideas.* Washington, DC: National Academies Press. *www.nap.edu/catalog/13165/a-framework-for-k-12-science-education-practices-crosscutting-concepts.*

Pearce, C. R. 1999. *Nurturing inquiry: Real science for the elementary classroom.* Portsmouth, NH: Heinemann.

Activity: Name That Object

Objective

To practice asking questions and use reasoning along with the senses of touch and sound to discover the identity of a hidden object

Materials

For each child: An opaque container with an easily closed lid (quart-size plastic yogurt containers or small cardboard boxes work well)

For each small group of students:

- A group of objects of various weights and surfaces, small enough to fit into the containers: a Ping-Pong ball, a plastic egg, various blocks, a large coin, a small paper cup, a feather, a cotton ball, and a marble. Use objects that are familiar to the students. *CAUTION: If children are younger than age 3, use only objects that are too large to swallow.*

- A tray to hold the objects

- A towel to cover the objects

Procedure

1. Show the children a tray of three to eight small objects and invite them to touch the objects, exploring their properties.

2. Ask each student to talk about one object and then compare two. *What do you feel? Is it soft or hard? How heavy is it? How are they alike, and how are they different?*

3. Play a "What is it?" game, in which students guess the identity of an object hidden inside the container. To keep students from seeing the object, cover the tray with the cloth while you pick up the object.

4. You can leave the cloth covering the remaining objects or remove it. Some children will ignore the objects left on the tray, but others will use them to determine what is not in the container. Acknowledge their use of this strategy by saying, *You are using a strategy that scientists also use. You asked the question, "What **can't** be inside the container? and answered by looking at what things were left on the tray. Scientists learn about the identity of an unknown thing by eliminating possibilities.* To make the game more challenging, after a few rounds keep the remaining objects covered.

5. For students who need guidance, model how to ask questions about what you feel and hear to determine what is inside the container. *Does it feel heavy? Does it roll or slide when I tip the container?*

6. The students will shake the containers to hear the sound the object makes as it moves. Encourage them to do slower movements, turning the container slowly side to side and top to bottom, to listen to the way the object moves and to feel its heft or weight. Ask, *What can we tell about the hidden object—which properties of the object can be discovered—by listening to it move? How heavy is it? If it feels heavy, can it be the feather? How does it move? Is it a roller or a slider?* A small paper cup and a Ping-Pong ball sound much alike and weigh about the same amount but can be distinguished by the way they move. On the other hand, two blocks with identical shapes but different colors cannot be differentiated. Give each child a turn to make a guess and share why they think that *before* opening the container.

7. Switch roles with the students, allowing them to put an object into the container while you cover your eyes. Keep the impetus for asking questions on the students by saying, *What questions should I be asking to learn about the hidden object?*

Many children will, in the first round of the game, guess their favorite object, usually the ball or coin. Don't correct even improbable answers. Instead, encourage thoughtful questioning about the properties of the hidden object by listening to all guesses and modeling the asking of questions. They will soon see if their guess is correct or not and, in further tries, will use their reasoning.

Teacher's Picks[9]

Publications

Bzzz, Bzzz! Mosquitoes in Your Backyard by Nancy Loewen (Picture Window Books, 2005).

Students learn more about these common insects as they read all about the life cycle of a mosquito in simple language. The illustrations are clear and nonthreatening for young children and help them understand the importance of mosquitoes in the balance of nature. The book is also a good choice for older kids because the back section includes "Fun Facts," "Words to Know," and the opportunity to learn how to "Drink like a Mosquito"!

Killer Bees by Elaine Landau (Enslow, 2003).

This book is perfect for primary students who love somewhat scary and unusual facts. It immediately captures the reader by describing a true story of a bee encounter, and it fascinates the reader with true stories intertwined with facts.

The Icky Bug Counting Book by Jerry Pallotta and Ralph Masiello (Charlesbridge, 1991).

Wonderful for prekindergarten through grade 2, this resource can be used to introduce interesting new "bugs" and teach about camouflage and counting; the book ends with a challenge question for the reader. It is a high-interest topic with simple text and a rhyming pattern that makes it a winner for integrating math and science.

Why Mosquitoes Buzz in People's Ears: A West African Tale by Verna Aardema, illustrated by Leo and Diane Dillon (Dial Books for Young Readers, 1975).

One of my favorite fiction books and a 1976 Caldecott Medal winner, this African legend has colorful illustrations and an amusing ending. It allows the teacher to integrate curriculum and connect with another culture, and it is a great complement to nonfiction insect books.

---

[9] These suggestions were provided by Mary Ann Hoffman, an ESL (English as a second language) teacher.

Websites

Classify Insects: Zoom In on True Bugs
http://teacher.scholastic.com/activities/explorations/bug/level1/investigate.htm

This Scholastic website illustrates the answers to questions about what kind of tools entomologists use to study insects. Students can click on the name of a tool to see a photo of it being used by an entomologist at work.

Science Online: Nature of Science
http://scienceonline.tki.org.nz/Nature-of-science

This webpage addresses the nature of science and related teaching activities. The webpage is part of Te Kete Ipurangi, and "online knowledge basket" that is an initiative of the New Zealand Ministry of Education.

Related *Early Years* Blog Posts

The following information is from "Garden Observations and Questions," published June 20, 2015:

> *Close observation can happen in a garden. Here are some questions that children can investigate:*
>
> - *Do all vines go around a pole in the same direction?*
>
> - *Why are there holes in this leaf? Is something eating it?*
>
> - *What happens to a leaf when rain falls on it?*
>
> *"Where are seeds made?" is a question to investigate over time by making observations of more than one plant. While drawing a plant in the garden, children may notice more than when walking through the space. A simple "journal" of a sheet of paper folded in quarters and a marker are all the materials needed to give children time to observe and think about what's happening in a garden.*

Read more at *http://nstacommunities.org/blog/2015/06/20/garden-observations-and-questions*.

Additional related blog post:

"When Are Children Old Enough to Smell a Flower, Touch an Earthworm, or Talk About the Nature of Science (NOS)?" published September 4, 2014: *http://nstacommunities.org/blog/2014/09/04/when-are-children-old-enough-to-smell-a-flower-touch-an-earthworm-or-talk-about-the-nature-of-science-nos*.

Chapter 19

Learning Measurement[10]

BEFORE YOU BEGIN THIS ACTIVITY:

Remember that each activity is not a stand-alone science or engineering curriculum. Activities are small steps in a journey of science inquiry, as discussed in Chapter 1. Your students will learn more about this concept and about the nature of science if you use this activity as part of an ongoing exploration of a question, a concept, or a topic being investigated by your class. Ask yourself, "What should come before this and what should come after?" Refer to Table 1.1 (pp. 15–39) to find other activities from The Early Years column that address the same concepts.

The world is filled with references to measurements that limit children's activities. It's no wonder children strive to be "big" when they often hear adults remarking, "That's too big a piece of cake for you," "That's too far for you to walk," "You're not old enough," and "You have to weigh 60 pounds before you may use a seat belt without a booster seat." Children may not have a set goal in mind, but they are delighted the first time they find themselves eye to eye with an adult. As they grow, their understanding of the concept of measuring also grows as they have experiences that teach them the meaning behind the adults' comments.

One situation that can confuse children about size is viewing photographs of creatures that have been enlarged to show details. The seemingly huge spider or earthworm is easier to contemplate when a coin is included in the photograph to show the scale. Coins and hands are measurement tools that bridge from early "play" experiences to lessons teaching about measuring with standardized measuring tools. By comparing an object such as a block with an

---

[10] This column entry was originally published in *Science and Children* in October 2006.

equal length in coins placed end to end, students begin to understand that the numbers on a ruler have meaning.

Volume can be particularly challenging to teach. Following a recipe is one way to teach measuring volume; another is the following reusable (and dry) activity. Cut off the bottoms and pointed tops of two clear plastic soda bottles (1 liter and 2 liters) and trim them to make two cylinders with identical volumes but different shapes: one tall and narrow and the other short and broad. To do this, hold one cylinder on a tray, partially fill with a volume of small dried beans, and mark the level they reach. Then pour that same volume of beans into the other column and mark the height. Trim the cylinders at the marks to get identical volumes. Ask children to point out the "bigger" of the two cylinders when empty. It's interesting that most will choose the tall one, showing a familiarity with measuring height. By pouring the beans from one cylinder to the other, students will accept what is not immediately obvious—that the cylinders have identical volumes.

Measuring is part of the science and engineering practice Using Mathematics and Computational Thinking identified in *A Framework for K–12 Science Education* (NRC 2012). The National Council of Teachers of Mathematics has specific measurement standards for prekindergarten through grade 2 (NCTM 2000). Attributes that can be measured—length, volume, weight, area, and time—are introduced in early childhood play with blocks, the water table, simple scales, fabric squares, and timers. Experiences in comparing two similar but not identical objects such as blocks or bowls lay the foundation for precise measurements in later lessons.

References

National Council of Teachers of Mathematics (NCTM). 2000. Principles and standards for school mathematics. *www.nctm.org/Standards-and-Positions/Principles-and-Standards/Measurement.*

National Research Council (NRC). 2012. *A framework for K–12 science education: Practices, crosscutting concepts, and core ideas.* Washington, DC: National Academies Press. *www.nap.edu/catalog/13165/a-framework-for-k-12-science-education-practices-crosscutting-concepts.*

Activity: Measuring Hands

Objective
To introduce measuring as a tool of scientists

Materials
- Construction paper

- Pencils

- Tape

- Clear contact paper or lamination (optional)

Procedure

1. Talk with students about size, asking for examples from their lives of objects that are large, medium, and small. Record the examples in lists by drawing or writing so students can see that some words, such as *dog, pizza,* or *bug* appear in more than one size category. Allow them to start new categories, such as *huge.*

2. Ask students to show with their hands the height of a big dog and then the size (diameter) of a large pizza. Depending on their experiences, they may discover that what is a big dog to one person is not necessarily a big dog to another, but pizza sizes are more standardized. Ask the students, *What if your family ordered a large pizza and the restaurant's large size was only as big as this (small) plate?* Ask the class to suggest other situations where being able to communicate size is important, such as cutting a doorway for a big dog.

3. Tell your class that as a group you will be making a tool to measure how long things are (length). Have the students trace around one hand with a pencil on construction paper. Encourage them to hold the fingers together to make the hand shape easier to cut out and to reduce tearing in later use. Students who finish quickly can make additional hands or decorate theirs. When they are finished, have them compare their hand length to objects in the classroom.

4. Collect the hands and stick each one to a long piece of tape, with the fingertips just touching the bottom of the previous palm so that the hands lie in a long continuous line. Strengthen this "measuring hands" tool by laminating it or applying a long piece of clear packing tape to each side.

5. Use the measuring hands tool to compare lengths. How tall is the doorway? How wide? How long is the rug, a desk, or a pencil? Sometimes short children are reluctant to compare heights since taller height is associated

with being grown-up and therefore more competent. If size is a sensitive issue in your classroom, or competition is too great for measuring body size, try comparing the distance around children's heads; that measurement is almost always very similar, differing from person to person by just an inch or two. With more use, students will realize that the measuring hands tool is useful for larger measurements but doesn't give enough information to compare objects smaller than a single hand. Once they come to that realization, it is time for them to begin working with standard measuring tools such as a 30 cm or 12 in. ruler.

Teacher's Picks

Publications

How Tall, How Short, How Far Away by David A. Adler, illustrated by Nancy Tobin (Holiday House, 1999).

This book is a good introduction to the development of measuring systems and standards for children who have had experience in measuring.

Measuring Penny by Loreen Leedy (Henry Holt, 1997).

Read this book to begin discussion about measuring different attributes. Dog owner Lisa finds out that there are many things about a dog to measure when she works on a school assignment to measure something in as many ways as you can. The text refers to both English customary units and metric units.

Capacity and *Length* by Henry Pluckrose (Children's Press, 1995).

These two books are part of the *Math Counts* series, which is a good resource for early readers because of clear text and photographs.

How Big Is a Foot? by Rolf Myller (Yearling, 1991).

This classic book is a humorous introduction to why measuring units are standard-ized. The story, measuring and building a bed for the Queen's birthday present, is adapted into a play in a lesson plan on the Utah Education Network website (see "Websites" section on p. 115).

Websites

Count Us In
www.abc.net.au/countusin/default.htm

This Australian Broadcasting Corporation webpage has self-correcting online games to help children explore a variety of math concepts, from counting to estimating capacity, without becoming frustrated. The illustrations depict both girls and boys with a variety of skin tones and clothing.

Path to Math: Measurement With Young Children
www.illinoisearlylearning.org/tipsheets/measure.htm

This is one of the online Tip Sheets from the Illinois Early Learning Project. Path to Math lists many ideas for including measurement in the daily rhythm of the early childhood classroom.

How Big Is a Foot?
www.uen.org/Lessonplan/preview.cgi?LPid=10729

This first-grade activity is part of a lesson plan from the Utah Education Network. The activity introduces nonstandard measurement and includes a worksheet to practice with estimating and measuring lengths, as well as a script for a play based on the book *How Big Is a Foot?* by Rolf Myller.

Related *Early Years* Blog Posts

The following information is from "Preparing the Classroom and School Grounds for Science Exploration," published August 29, 2013:

Magnifiers are a tool that can enhance children's explorations of size. Two-year-olds can learn how to hold magnifying glasses to get a close-up look at leaves, feathers, and other interesting objects. Make these tools available throughout the school

- *in the dress-up area for imaginative play or looking closely at fabrics,*

- *in the book corner for examining illustrations,*

- *at snack time to see the wrinkles in a raisin or the fibers of a celery stick,*

- *outside on the playground or in a teacher's pocket to be available to look at small creatures or grass leaves, and*

- *at the fish tank to see the details of a fish's body.*

Read more at *http://nstacommunities.org/blog/2013/08/29/preparing-the-classroom-and-school-grounds-for-science-exploration.*

Additional related blog post:

"Sensory Table Explorations of Matter," published December 19, 2013: *http://nstacommunities.org/blog/2013/12/19/sensory-table-explorations-of-matter.*

Chapter 20

Rocks Tell a Story[11]

BEFORE YOU BEGIN THIS ACTIVITY:

Remember that each activity is not a stand-alone science or engineering curriculum. Activities are small steps in a journey of science inquiry, as discussed in Chapter 1. Your students will learn more about this concept and about the nature of science if you use this activity as part of an ongoing exploration of a question, a concept, or a topic being investigated by your class. Ask yourself, "What should come before this and what should come after?" Refer to Table 1.1 (pp. 15–39) to find other activities from The Early Years column that address the same concepts.

Sedimentary rocks, formed by an accumulation of sediments (tiny pieces of rocks or minerals) in a water environment, tell a story that many students may be familiar with. They may have visited areas where water or wind carried sediments and deposited them in rivers, lakes, oceans, or dunes. The rocks are often visually or texturally interesting and may have the added attraction of containing fossils.

We can understand the stories rocks tell more easily if we have experience with the materials that make up a rock. Here are a few suggestions on how to experience these rock materials:

- With permission, if needed, dig clay or sand from the ground to bring back to the classroom. Examine it, and wash a cupful in water on a tray to see what else is in the sample—perhaps "dirt," organic matter from plants, small pebbles, and shells.

---

[11] This column entry was originally published in *Science and Children* in December 2006.

- Take a field trip to a beach on an ocean, lake, or river to see sediments accumulate.

- Add a small amount of sand or clay to standard paints for painting pictures on paper.

- Pour water into a tub of sand to see how it can move sand. *CAUTION: When finished, do not dispose of the sand in the sink drain.*

- Mix sand and clay with water in jars to shake and watch the sediment settle. Make one jar with ¼ cup sand, one with ¼ cup clay, and one with 2 tablespoons of each, and seal tightly with hot glue inside the lids and tape outside. Ask questions before shaking: *What do you think will happen if we mix the sand or clay with water? How long will it take the sand to mix into the water, and how long will it take the clay? What will happen when you stop shaking? What did you find out?*

When the clay and sand that the children have been working with dry out, the children will notice that they no longer stick together. The sand is once again individual grains and the clay, although it's hard to see particles, feels "dusty" and is easily broken. *CAUTION: Keep wet paper towels nearby to clean up dry clay dust rather than sweeping it up, to avoid airborne dust (Roy 2014).*

As part of the exploration of earth materials, students can record their ideas about why clay and sand feel different and how they hold together when wet and when dry. Direct exploration of clay and sand builds students' understanding of their properties. Describing and classifying materials by their observable properties is part of the grade 2 performance expectation 2-PS1-1 in the *Next Generation Science Standards* (*NGSS*), and experiences with clay and sand can help students understand Earth events, which is another grade 2 performance expectation (2-ESS1-1) in the *NGSS* (NGSS Lead States 2013).

In the following activity it is very important to always use the term *pretend rock* so children do not get the idea that rocks are human made. Rocks are formed through natural processes. Many descriptions of rock formation for young children say that the rock formed when sediment was buried under tons of more sediment and dirt until it turned into solid rock, omitting the role of cementing materials. Sediments are cemented together when water carrying dissolved minerals seeps into the spaces between the particles and the minerals precipitate out from the water in the spaces, cementing the grains together. In the Pretend Rocks activity, plaster of paris will be added to the sediments in the cup to act as the cement.

Variation seen in sedimentary rocks comes from many differences, including color of parent material, source of parent material, particle shape, particle size, and the environment in which the sediment was deposited. You can offer various sediments in this activity to produce a variety of pretend rocks.

References

NGSS Lead States. 2013. *Next Generation Science Standards: For states, by states.* Washington, DC: National Academies Press. *www.nextgenscience.org/next-generation-science-standards.*

Roy, K. 2014. Safety first: Modeling safety in clay use. *Science and Children* 50 (4): 84–85.

Activity: Pretend Rocks

Objective

To notice the range in grain size in sedimentary rocks and think about how such rocks are formed

Materials

- Samples of various sedimentary rocks, including sandstone and shale, made of different-size particles (Rock can be purchased through local stone dealers or scientific supply companies or collected locally.)

- Magnifier

- Rock identification book for general audience

- Sand

- Ceramic (pottery) clay formulated for safe use by children (see "Resources" section)

- Spoons to serve the sand and clay

- Paper towels

- One 5–8 oz. paper cup and craft stick for each child

- Plaster of paris

- Disposable containers to mix plaster

- Water

- Pebbles, dirt, and small shells (optional)

Procedure

1. As the collected rocks are brought in, a discussion about where rocks come from develops. Ask if anyone has ever seen a rock being made, to learn the students' ideas. Encourage the children to look closely at the rocks and compare them with each other in color, size, shape, texture, and weight. Have them use a magnifier to see the grain size. *What size are the pieces that make up these rocks? Can you see them or feel them? Are any of these rocks the same?* Note that a rock can be many sizes and have different names—such as sand, pebble, stone, and boulder—and still be rock.

2. Compare the actual rocks with those pictured in a rock identification book. The "match" that young children make is usually based on color and shape rather than other distinguishing properties or origin. At this age they are beginning to understand the use of an identification book, not the complexities of rock composition, so no corrections are needed.

3. Tell the students that now they are going to feel the raw materials that make sandstones and shales. Have the students feel soft, damp clay and damp sand, keeping the materials separate. Accentuate the difference in textures by using clay that does not contain grit so its texture is very smooth. *Where have you seen clay or sand? How are these two materials alike or different? What size are the pieces that make up the clay and sand?*

4. Ask for ideas on how to make a "pretend rock." Then tell the children that you have a recipe to try. Give each child a small paper cup to fill about half-full with damp sand or very wet clay. You might also provide pebbles, dirt, or shells to be added. Make one pretend rock of just clay and one of just sand so the children can later compare these types of pretend rocks. Have the children stir their chosen material(s) using a craft stick (it doesn't accidentally flip sand the way a spoon does).

5. Using a finger, test to see if the mixture "is a rock yet." Tell the children that a "cementing" material must be added, then add a heaping teaspoon of mixed plaster of paris. *CAUTION: Only adults should mix and add the plaster of paris; follow the package instructions, and do not wash the remainder down the sink.* Have the children stir their mixture thoroughly and describe it.

6. Review the process for sedimentary rocks formed by an accumulation of sediment: *Rock formation is happening all the time, not in schools or factories but in nature. It takes a long time for the sand grains and clay minerals to pile up in the same place and become buried as more sediment is deposited on top, and*

for water to carry dissolved minerals into the sand and clay minerals to become the glue that holds the pieces together. By tomorrow—a much shorter time—our pretend rock materials will be cemented and become hard.

7. After 24 hours, have the students peel off the paper cup to reveal the pretend sedimentary rock. Doing this as a group will allow the children to compare rocks and talk about how their rocks are made of different-size particles.

As a follow-up to the activity, make a snack "rock" with a variety of cereal particle sizes, including the puffed rice cereal in the original Rice Krispies Treats recipe (*www.ricekrispies.com/recipes/the-original-treats*). The melted marshmallows are the cement!

Resources[12]

Art and Creative Materials Institute, Inc. (ACMI). 2015. Safety tips—what you need to know. *http://acminet.org/index.php?option=com_safetytips&view=safetytips&Itemid=30.*

Gyllenhaal, E. 2002. Aaron's treasures. *Chicago Parent,* July. Also available at *http://saltthesandbox.org/ChicagoParentArticle1.htm.* (This article was written about preschool children's passion for collecting.)

Gyllenhaal, E. 2001–2002. Neighborhood rocks. *www.saltthesandbox.org/rocks/index.htm.* (This website has lots of ideas for finding rocks in cities and suburbs and for collecting, identifying, and playing with these finds.)

Gyllenhaal, E. 2001–2009. Salt the sandbox. *http://saltthesandbox.org.* (This collection of websites is devoted to the interests of young children, especially topics related to nature and science.)

Teacher's Picks[13]

Publications

A Gift From the Sea by Kate Banks, illustrated by George Hallensleben (Farrar, Straus and Giroux, 2001).

---

[12] Eric Gyllenhaal, a geologist and museum educator and evaluator, developed the three resources on Earth science and collections; they are useful for parents and teachers of young children.

[13] These suggestions were provided by Marie Faust Evitt, a preschool teacher and author of the book *Thinking BIG, Learning BIG: Connecting Science, Math, Literacy, and Language in Early Childhood* (Gryphon House, 2009). The book's website has a useful "Links/Resources" tab: *http://thinkingbiglearningbig.com.*

The sumptuous illustrations and lyrical text describe the journey a rock takes from a volcano through the Ice Age and early civilization to the bottom of the ocean and finally to the beach where a boy finds it. Though the text is simple, you can use it with older children to help them speculate about the history of rocks they find.

Rocks in His Head by Carol Otis Hurst, illustrated by James Stevenson (Greenwillow Books, 2001).

Understated humor punctuates this true story about the author's father, whose passion for rock collecting as a boy eventually leads him to become curator of mineralogy at a science museum. Learning about this natural-born scientist researching, labeling, and displaying his beloved rocks will inspire students to follow their dreams.

Grand Canyon: A Trail Through Time by Linda Vieira (Walker, 1997)

This rich description of one of the natural wonders of the world reads as a story. Facts about the canyon's record of geologic time are interwoven with information about the animals, plants, and people who live in it today.

The Sun, the Wind, and the Rain by Lisa Westberg Peters, illustrated by Ted Rand (Henry Holt, 1988)

This beautifully illustrated book provides an excellent introduction to geologic processes by comparing the creation and evolution of mountains with a sand hill that a girl builds at the beach. Children can readily see the connection between their own experiences with sand and the weathering of the natural landscape.

Websites

Geology of National Parks, 3D and Photographic Tours; Geology of the National Parks: Virtual Tours
http://3dparks.wr.usgs.gov; http://geomaps.wr.usgs.gov/parks/project/index.html

From rock formations at Bryce Canyon National Park to the stones used to build our nation's capital in Washington, D.C., these U.S. Geological Survey websites present the outstanding geology of many different parks.

Images of Clay
www.clays.org/EDUCATIONAL%20RESOURCES/ERimages.html

This website, a joint initiative of the Clay Minerals Society and The Clay Minerals Group, provides a look at highly magnified photos of different clay minerals that will help children understand that clay, like sand, is made of particles.

Related *Early Years* Blog Posts

The following information is from "Rocks: Collecting and Classifying," published June 13, 2009:

> *Walking along a creek is one place to find rocks that have been moved there by natural forces, not by humans. (Be sure to wash hands afterward.) You don't have to know what type of rock it is to appreciate that it is smooth and pinkish, or has sparkles, or has holes in it.*
>
> *Label even the most nondescript rock with the location and date collected, and that single rock becomes the beginning of a scientific rock collection. A high school Earth science teacher might be willing to view the collection and help with scientific names.*

Read more at *http://nstacommunities.org/blog/2009/06/13/rocks-collecting-and-classifying*.

Additional related blog post:

"Exploring Natural and Human-Made Materials," published December 20, 2012: *http://nstacommunities.org/blog/2012/12/20/exploring-natural-and-human-made-materials*.

Chapter 21

Birds in Winter[14]

BEFORE YOU BEGIN THIS ACTIVITY:

Remember that each activity is not a stand-alone science or engineering curriculum. Activities are small steps in a journey of science inquiry, as discussed in Chapter 1. Your students will learn more about this concept and about the nature of science if you use this activity as part of an ongoing exploration of a question, a concept, or a topic being investigated by your class. Ask yourself, "What should come before this and what should come after?" Refer to Table 1.1 (pp. 15–39) to find other activities from The Early Years column that address the same concepts.

Science and art go hand in hand, because many scientists make art to share their observations. Two scientists who are widely known first as artists are Leonardo da Vinci, anatomist and inventor, and Beatrix Potter, mycologist. Both of these scientists used observations of the natural world as the springboard for their discoveries.

Young children like to paint and draw and are learning that they can make representations and share their stories with artwork. Use your students' interest in drawing to teach about recording and sharing scientific observations. Making observations about living things is part of kindergarten performance expectation K-LS1-1: "Use observations to describe patterns of what plants and animals (including humans) need to survive" (NGSS Lead States 2013).

Birds are an interesting subject to observe and draw because they are so beautiful and so varied. Children can observe birds, draw birds, identify birds, and measure and count them. Making observations gives an additional purpose to taking the class on a quick walk around the school building whenever they need a short break to sharpen their attention. Comment on birds around the school and passing by the playground. Name them or ask the children to name them, if possible. It's okay to use an appropriate made-up name such as "the black ones with the white speckles" or the "red bird" to help others identify the bird another time. The birds you see will vary depending on many factors, including place on the continent, elevation, proximity to water, amount of vegetation nearby, time of year, and time of day.

---

[14] This column entry was originally published in *Science and Children* in February 2007.

For several days to a week, have your students observe birds during a walk around the school or a few minutes of their recess and record what they see by answering these questions: *Where do you see birds? Are the birds in trees, on the ground, in the water, or in the air? Do you see the birds in the same place every day? What are they doing? Where have you seen birds eating? What size (color, shape) are the birds?* The questions can be written on a checklist for students who read to record one- or two-word answers. The answers for that day or week can be tallied and graphed. With subsequent observations, students will begin to recognize some birds and describe them by color, relative size, and sometimes habit—where they are found or what they are doing.

In some regions it makes sense to do this activity in winter because there may be fewer birds than usual due to bird population migration—and they can be easier to spot when trees are leafless. In regions where winter temperatures may not allow for outdoor observations, do this activity in another season.

Reference

NGSS Lead States. 2013. *Next Generation Science Standards: For states, by states.* Washington, DC: National Academies Press. *www.nextgenscience.org/next-generation-science-standards.*

Activity: Bird Shapes

Objective

To identify birds common to your area by shape and color and to introduce scientific illustration

Materials

- Bird identification chart or books

- Sheets of stiff material, such a noncorrugated cardboard from cereal boxes

- Large crayons with the labels peeled off in colors that commonly occur in birds (typically black, white, red, blue, gray, and brown)

- Paper

- Masking tape

Teacher Preparation

Make silhouette shapes of local birds the class has identified by tracing around the pictures in an identification book and using a copier or overhead projector to blow the image up to life size (see measurements listed in the book). Include birds of various sizes and different colors. Cut the shapes from sheets of stiff material such as noncorrugated cardboard from cereal boxes.

To make a rubbing, tape each shape to a table by sticking loops of masking tape on the back, then tape a piece of paper over the shape to cover it. Using the appropriate color(s), rub or "wipe" the paper with crayons. You can add minimal details such as eyes, legs, and stripes to make the bird easier to identify.

Procedure

1. Have the students use an identification chart or books to identify birds they have seen. Some birds seem the same at first glance—a chickadee might be mistaken for a sparrow because of its small size—but by focusing on the predominant color, students can usually tell them apart.

2. While looking in the chart or books, comment that by drawing exactly what the birds look like in nature, the artist shows other people what to expect—what the actual birds look like. Introduce the idea that bird species are in part defined by their colors by saying, *The small brown bird is a sparrow and is always brown and is never blue. If you see a small blue bird, it must be a different species of bird.*

3. Give the children the bird shapes to handle and ask them to identify "their" bird by comparing its shape to the teacher-made rubbing examples (posted on the wall) or pictures in an identification chart or books. Ask open-ended questions to help them narrow the choices: *Is the bird big or little? What shape is the head? What kind of tail shape does the bird have?* Children can test their match by laying the shape on top of the rubbing examples. Although at first glance a student might mistake a chickadee for a sparrow, when questioned about beak size, they usually notice the difference.

4. Help the children tape the bird shapes to a table, tape paper over the shapes, and feel the shape through the paper. Then they can make their own rubbings using appropriate colors. To help children understand the reason for using appropriate colors, say, *If you want to tell me about a bird that is brown, you have to use brown to make the rubbing or else I might go looking for a pink bird instead of a brown one.* Explain that by using the actual colors of the birds in our rubbings, we can show other people what the real birds look

like. Scientists record what they see in nature. We could look for a long time and would never see a pink crow in nature! This does not mean that the children should limit their purely creative artwork to nature's colors.

5. Have the students "feel" the paper with the crayon to make the rubbing. Children grasp the idea of rubbings better if you tell them to "pinch" the crayon sideways and to "wipe" or "feel" the paper with the crayon. If this technique is new to them, children are astonished to see the bird shape emerging with each pass of the crayon. Encourage them to add details such as eyes and feet.

Extension

Make additional rubbings of birds for use in tallying the number of times a bird species is seen. Tape the rubbings vertically on a poster to create a graph of how many of each species were sighted during a period of time. Graphs make it easy to see how many of each species of bird the children observed. Adding the bird names to the graph is optional.

To emphasize that science and art are both important human activities, make the next project about birds one in which the children create a new species, perhaps with wild colors but a familiar shape.

Teacher's Picks

Publications

Kaufman Field Guide to Birds of North America by Kenn Kaufman (Houghton Mifflin, 2005).

A field guide with digitally enhanced photographs and pointers (leader lines) to field marks (distinguishing markings or aspects to note that aid identification), this book is fun for children to use for identification purposes, to browse through, or to read about a particular species.

Counting Is for the Birds by Frank Mazzola Jr. (Charlesbridge, 1997).

Useful as both a counting book and a bird identification book, sections of this informative book can also be read aloud as children point out birds they have seen.

How the Robin Got Its Red Breast: A Legend of the Sechelt People by the Sechelt Nation, illustrated by Charles Craigan (Nightwood Editions, 1993).

Compelling black-and-white illustrations relate the Sechelt legend of how the robin got its colors. Traditional tales often teach about the natural world while serving as guides for proper conduct.

Rainbow Crow by Nancy Van Laan, illustrated by Beatriz Vidal (Dragonfly Books, 1991).

A retelling of a Lenape Indian story tells how the crow became black, a result of his bravery. Several species of birds are illustrated, all in natural colors.

Websites

BirdWatching
www.birdwatchingdaily.com/getting-started

Get answers to "What to do if you find a baby bird?" and learn the basics of bird ID in the "Getting Started" section. "Photo of the Week" offers readers' views of many bird species.

All About Birds
www.birds.cornell.edu/AllAboutBirds/birding123/identify

On the "How to ID Birds" section of this website from the Cornell Lab of Ornithology, learn how to use the following features as aids to bird identification: silhouette, field marks, posture, size, flight pattern, and habitat. The silhouettes in these sections are particularly helpful in comparing bird size.

Related *Early Years* Blog Posts

The following information is from "Kindergarten Teacher Shares Her Class' Bird Investigation," published January 23, 2014:

> Guest blogger Mary Myron, an experienced lead teacher in early childhood programs, discussed her Bird Project:
>
> It has been to date the longest and most in-depth project I have been involved in. ... It was such an amazing experience that I love to share it. For curriculum planning purposes, I use an overall umbrella theme for a period of weeks or even months. It is always a science-related theme and

usually has to do with the changes that are occurring out of doors in our northeastern Tennessee environment. I select these themes because they are meaningful and relevant to young children as curious scientists.

Read more at *http://nstacommunities.org/blog/2014/01/23/kindergarten-teacher-shares-her-class-bird-investigation.*

Additional related blog posts:

"Recording Observations and Collecting Data About Birds," published March 2, 2011: *http://nstacommunities.org/blog/2011/03/02/recording-observations-and-collecting-data-about-birds.*

"Searching for Evidence of Animals Using Plants for Food or Shelter," published May 28, 2013: *http://nstacommunities.org/blog/2013/05/28/searching-for-evidence-of-animals-using-plants-for-food-or-shelter.*

Chapter 22

The Sun's Energy[15]

BEFORE YOU BEGIN THIS ACTIVITY:

Remember that each activity is not a stand-alone science or engineering curriculum. Activities are small steps in a journey of science inquiry, as discussed in Chapter 1. Your students will learn more about this concept and about the nature of science if you use this activity as part of an ongoing exploration of a question, a concept, or a topic being investigated by your class. Ask yourself, "What should come before this and what should come after?" Refer to Table 1.1 (pp. 15–39) to find other activities from The Early Years *column that address the same concepts.*

Understanding the connection between the Sun's energy and sustaining life is difficult for preschoolers, but learning about these concepts through both long- and short-term activities captures children's short attention spans. Activities such as growing plants in sunlight and without light, playing with light and shadow, and making "Sun prints" explore light—in this case, how the Sun's light is different from lamplight.

The concept of how energy from the Sun supports life on Earth is a complex idea. The book *Brown Cow Green Grass Yellow Mellow Sun* (Jackson 1995) introduces this in an appropriate way for young children, telling a simple story of grass using the Sun's energy to grow, a cow eating the grass and growing, and people drinking the cow's milk. Although children's grasp of these concepts will vary, having direct experiences with a variety of materials and being encouraged to think about what they experience and observe is a foundation that supports children's later learning. Activities about light and sunlight address the *Next Generation Science Standards* kindergarten performance expectation K-PS3 Energy, as well as the disciplinary core idea PS3.B: Sunlight warms Earth's surface (NGSS Lead States 2013).

Outdoors on a sunny day, begin a discussion about sunlight with the familiar subject of our senses. *CAUTION: Remind young children repeatedly that they should not look directly at the Sun to avoid damaging their eyes.* Ask your students, *How can we sense the sunlight? Can we taste (hear, smell, see, feel) it?* Pretend to taste the sunlight and sniff it. This unusual behavior will focus students' attention on something they

---

[15] This column entry was originally published in *Science and Children* in March 2007.

have experience with but probably never thought about: "What is light?" With great expression, feel the sunshine on your skin and ask your students to tell you how it feels to them. Depending on the climate and season, children may answer, "Good!" or "Too hot." Ask students to relate what happens if they get too much sunshine on their skin and if other light sources have the same effect.

Engage the students in a discussion of how sunlight helps us. Aside from providing energy to plants, sunlight warms the Earth and allows us to see. Ask, *How is the light from the Sun similar to the light from a lamp?* Both are necessary for us to see. *What would happen at night if we didn't have lamps? How is the light from a lamp different from the sunlight?* This discussion gets students thinking about different types of energy. Prepare the materials for the following activity ahead of time so your class can take advantage of a sunny day and use the Sun's energy to make art.

References

Jackson, E. 1995. *Brown cow green grass yellow mellow sun.* New York: Hyperion.

NGSS Lead States. 2013. *Next Generation Science Standards: For states, by states.* Washington, DC: National Academies Press. *www.nextgenscience.org/next-generation-science-standards.*

Activity: Making Sun Prints

Objective

To understand that although we can't see the ultraviolet light part of sunlight, it can have an effect on plants, our skin, and light-sensitive chemicals

Materials

- Sun-reactive paper kit, available from craft supply stores and science suppliers (This paper has a chemical coating that reacts in ultraviolet light to change color.)

- Flat objects such as leaves, shapes cut from aluminum foil, or opaque tracing templates (Flat objects cover the paper more thoroughly to make a more distinct picture.)

- Plexiglas sheet to hold the paper in place (Inexpensive poster frames are a good source, but this is optional—holding the paper flat makes a more well-defined print.)

- Pan of water large enough to wash the paper in

- A book to read aloud for about five minutes

Teacher Preparation

In advance, make an outdoor sun print to use as an example of a well-defined print. Follow the kit instructions if they are different from the instructions below. Make a well-defined, high-contrast print by keeping the paper with the object in the sun-light for five minutes.

Procedure

1. Make an indoor sun print as a rehearsal for making the prints outside. Place a flat object on top of the paper (colored side up), cover the paper and object with a Plexiglas sheet, and leave it in the light inside for at least 20 minutes before stopping the chemical reaction by gently washing the paper in a pan of water. As the paper dries, the color becomes brighter.

2. Show students both the outdoor and the indoor prints and compare them. Ask, *Where do you think the brighter one was made?* and *Why is the color of the one I made outside so much brighter than the one we made inside?* Students may think of the Sun if they are asked to think of the brightest light they know.

3. Before making the prints outdoors, choose a short book to read outside while waiting for the images to form. Then begin the process of making prints outdoors by setting up a pan of water in the shade and setting up the Plexiglas covering for the sun prints in a sunny spot. The chemical coating on the paper begins reacting immediately! Work with a group of six or fewer students at a time, and have students choose their objects ahead of time. Leaves or a pattern of small stones make beautiful silhouettes. Quickly write each student's name on the sun-reactive paper as you remove it from the package, or attach a sticker prepared with the name written on it.

4. Instruct students to place their object on the paper. Covering the paper with a sheet of Plexiglas keeps everything from blowing away while the print develops, or they can weigh it down with pennies. Describing what is going to happen—*I'm going to put Saul's paper here, and he is going to quickly put his objects onto the paper*—helps students work more quickly.

5. While waiting for the images to form (about five minutes), move into the shade and read a short book about light. Ask open-ended questions about the Sun and how its energy is the source of energy for almost all life. *What do we need to grow and where do we get it? What do plants need to grow? Where do they get it?* The Sun's energy is making the special paper change color.

6. After five minutes, have the children quickly pick up their papers and put them into the pan of water and step back to allow others to do the same.

7. Gently swish the papers around to wash the chemicals off the paper to stop the reaction. Spread the papers out individually to dry.

8. When the prints are dry, the students can compare them to the one made inside. Understanding is demonstrated when children attribute the change in the paper's color to exposure to sunlight.

Extension

Extend understanding of sunlight the next time the class eats together by remarking how the food is made from plants—plants that used energy from the Sun to grow. The next time the class goes outside on a sunny day, take a prism with you to show the class that the visible part of sunlight can be separated into a rainbow of colors. With these experiences under their belts, students will begin making connections between the Sun's energy and life on Earth.

Teacher's Picks[16]

Publications

A Rainbow All Around Me by Sandra L. Pinkney, photographs by Myles C. Pinkney (Scholastic, 2002).

Bold close-up photos of children from a rainbow of ethnic groups expand exploration of color into skin color. The book invites children to reflect on the many moods, emotions, and sensations evoked by color, including nonrainbow colors like pink, tan, and brown. It concludes, "Colors! They're in everything I see! We are the rainbow—YOU and ME!"

The Rainbow and You by E. C. Krupp, illustrated by R. R. Krupp (HarperCollins, 2000).

The character Roy G. Biv leads children through the scientific explanations of rainbow formation. The author includes brief summaries of global rainbow legends from the ancient Greeks to the American Navajo.

---

[16] These suggestions were provided by Marie Faust Evitt, a preschool teacher and author of the book *Thinking BIG, Learning BIG: Connecting Science, Math, Literacy, and Language in Early Childhood* (Gryphon House, 2009). The book's website has a useful "Links/Resources" tab: *http://thinkingbiglearningbig.com*.

All the Colors of the Rainbow by Allan Fowler (Children's Press, 1998).

Even complicated concepts are understandable in this simple text about how rainbows are formed: "Mixing light is similar to mixing paint. If you mix red and yellow, you get orange. So orange lies between red and yellow in a rainbow."

Website

What Are Those Squiggly Lines? Using Light to Learn About the Universe
http://violet.pha.jhu.edu/~wpb/spectroscopy/spec_home.html

This website from Johns Hopkins University is a good resource for teachers on the concepts of light. See "The Basics of Light" page to familiarize yourself with the concept of light as energy and the "The Electromagnetic Spectrum" page to see a graphic showing that visible light waves are just a portion of the entire electromagnetic spectrum.

Related *Early Years* Blog Posts

The following information is from "Observing Weather Events," published January 3, 2013:

> What interesting natural events have you noticed that occur each year in your area or schoolyard? Young children notice some changes, but others happen little by little and are not noted. Documenting the gradual changes of hours of sunlight, leaf color changes or leaf drop, windiness, or cloud cover can bring these changes to children's attention.

Read more at *http://nstacommunities.org/blog/2013/01/03/observing-weather-events*.

Additional related blog post:

"Heat and Energy: What Can Young Children Understand?" published February 9, 2015: *http://nstacommunities.org/blog/2015/02/09/heat-and-energy-what-can-young-children-understand*.

Chapter 23

Collards and Caterpillars[17]

BEFORE YOU BEGIN THIS ACTIVITY:

Remember that each activity is not a stand-alone science or engineering curriculum. Activities are small steps in a journey of science inquiry, as discussed in Chapter 1. Your students will learn more about this concept and about the nature of science if you use this activity as part of an ongoing exploration of a question, a concept, or a topic being investigated by your class. Ask yourself, "What should come before this and what should come after?" Refer to Table 1.1 (pp. 15–39) to find other activities from The Early Years column that address the same concepts.

Community, *assemblage, network, complex, interdependent, web,* and *synergism*— definitions of an ecosystem often include these words to highlight the dynamic interrelated workings of plants and animals with their physical environment. Young children don't understand the complexities of ecosystems, but they can begin to understand that only certain food sources meet the needs of an insect species. Using observations to describe patterns of what plants need to survive is part of the *Next Generation Science Standards* kindergarten performance expectation K-LS1-1, From Molecules to Organisms: Structures and Processes (NGSS Lead States 2013).

The cabbage white butterfly (*Pieris rapae*) is common throughout much of North America. Food growers consider it a pest because its larvae eat leaves of the Brassicaceae (also known as Cruciferae) family, such as cabbage, broccoli, collards, and kale. But to the early childhood teacher, it is a convenient way to teach about ecosystems. In fall or spring, plant a few flowering annuals along with a row or pot of collard greens seedlings, and you will probably see these white-to-cream colored butterflies dancing around the plants within a week. The flat open

---

[17] This column entry was originally published in *Science and Children* in April 2007.

leaves of collards make it easier to find caterpillars and eggs. Later you may see holes in the leaves and find bright green caterpillars on the underside of the leaves. They are hard to see because they are almost the same color as the leaf and have few markings. The larvae hatched from tiny oval eggs, singly attached to the leaf by the female butterflies as they laid them. You may find both eggs and larvae on the plants because eggs are laid throughout the summer. *CAUTION: When venturing outside with students, even into a garden, check with parents for allergies and know the potential stinging or biting organisms in your area.*

Asking questions will focus students' attention on the interrelatedness aspect of ecosystems: *What do the butterflies eat? What do they drink? What other living things do they share food and space with? What do their babies eat? What eats them? Do their daily actions affect other living things as they interact?* Look for answers to these questions by observing the butterflies through all life stages. Ask your students, *What do you see in the collard patch today? Has anything changed since last time?* Observations such as, "I touched an egg on the leaf and it squished" or "Two butterflies are going around the collards" can be quickly dictated to an adult to record them in a notebook or documented with student drawings. Counting visible butterflies, eggs, larvae, and other animals or the number of holes in a leaf are ways to take note of animals in the ecosystem and to notice any changes. Ask, *If the number of caterpillars decreases, is it because some crawled away, or were they eaten by another animal?*

Over several weeks of making observations, students will find out that the needs of adult insects differ from the needs of the larvae. To see that the caterpillars will eat only some kinds of leaves, bring a few caterpillars into the classroom in a container. When they form pupae, the class can observe this intermediate life stage and the next—butterflies.

Reference

NGSS Lead States. 2013. *Next Generation Science Standards: For states, by states.* Washington, DC: National Academies Press. *www.nextgenscience.org/next-generation-science-standards.*

Activity: What Do Caterpillars Eat?

Objective
To notice the interrelatedness of the cabbage white butterfly and the Brassicaceae (or Cruciferae) family of plants

Materials
- An insect field guide

- Cabbage white butterfly (*Pieris rapae*) caterpillars

- A large container to house the caterpillars (wide-mouth jar or commercial "butterfly habitat" [zippered hamper])

- Cloth or paper towels for lid

- Rubber bands

- Plastic cup and lid

- Collard leaves (*Brassica oleracea* var. *acephala*)

- Leaves from a plant that is not in the Brassicaceae (or Cruciferae) family, such as grass or oak tree leaves

- Cotton balls to hold water

Procedure

1. Locate cabbage white butterfly larvae on the leaves of a brassica plant and observe them eating.

2. Brainstorm with the students about what the "butterfly babies" will need if they visit the classroom for a few weeks. The discussion usually reveals that they do not need parental care but do need food, water, and the "right" temperature. A few children may mention the need for air. Discuss that you will be taking the caterpillars out of their ecosystem and when even one small part of an ecosystem is removed, other parts will likely be affected. When deciding how many caterpillars to bring inside, ask, *What would happen if all the cabbage white caterpillars were removed from their ecosystem?* Although this species is far from endangered, it is good to practice having a minimal impact on any species studied.

3. With the children, set up an insect habitat for a few cabbage white caterpillars using a large clear plastic snack food jar or other wide-mouth container. Wash the container thoroughly. To allow air exchange, cover the opening with a cloth or paper towel held on with a rubber band.

4. Set up food for the caterpillars. For convenience, feed the caterpillars with a bouquet of small collard leaves set in a container with the stems in water. Have the students carefully wash the collard leaves before putting them in the container. A clear plastic cup with a carryout lid works well because the water level is visible and leaf stems can be inserted through the opening (but it's too small to allow caterpillars to fall into the water).

5. Put additional leaves such as grass or oak tree leaves in the container to see if the caterpillars will eat them.

6. Students can sprinkle a few drops of water into the habitat every day or put in a fresh wet cotton ball to provide a source of water.

7. Watch the caterpillars move, eat, and excrete. Record the number of holes in leaves, the number of leaves consumed, and what kind of leaf is being eaten. The larvae will pupate and form chrysalides within approximately two weeks. The change to pupae most likely won't happen when anyone is watching. The caterpillar will shed its skin and the new skin will look different.

8. As butterflies emerge from their chrysalides in about 10 days, release them near the collard patch.

After this introductory activity, children will eagerly find other caterpillars. *CAUTION: Remind children not to pick up caterpillars without asking an adult first, to protect children and caterpillars.* Remind them that the cabbage white caterpillars did not eat the "other" leaves, and ask, *What kind of leaf does this kind of caterpillar eat?* Be ready to search for the answer using field guides (see "Resource" section).

Children can eat caterpillar food, too! They enjoy washing, cutting, and seasoning the collard leaves and eating them after an adult does the cooking.

Resource

Wright, A. B. 1998. *Peterson first guide to caterpillars of North America.* New York: Houghton Mifflin.

Teacher's Picks[18]

These are good books to read to children about gardens and the ecosystems in them.

Publications

Sunflower House by Eve Bunting, illustrated by Kathryn Hewitt (Harcourt Brace, 1996).

A young boy and his father plant a sunflower playhouse for him and his friends to play in; it lasts for the summer, but then the flowers start to die. At first the children

---

[18] These suggestions were provided by Yvonne Fogelman, an educator whose interests included family science and math programs for preschoolers through second grade.

are sad and try to fix the sunflower playhouse, but then they realize that they can take the seeds and plant a new garden next year. As they stuff the seeds into their pockets, some seeds fall and birds swoop down to eat them. Use this to illustrate one route seeds travel.

Flower Garden by Eve Bunting, illustrated by Kathryn Hewitt (Harcourt Brace, 1994).

A young girl and her father buy flowers at the grocery store to plant into a windowsill garden as a birthday surprise for her mom. For the many children who don't live in houses with yards, this is a good resource to show that even apartment dwellers can have a garden.

Who Eats What? Food Chains and Food Webs by Patricia Lauber, illustrated by Holly Keller (Let's-Read-And-Find-Out Science, Stage 2; HarperCollins, 1994).

Have second graders draw their own food chains after reading the descriptions in this book, including one of a leaf-caterpillar-wren-hawk food chain.

Planting a Rainbow by Lois Ehlert (Harcourt, 1988).

A little girl and her mom plant flowers in a rainbow of colors. The illustrations show how different bulbs, corms, and rhizomes are planted at different depths.

Pumpkin Pumpkin by Jeanne Titherington (Greenwillow Books, 1986).

This is a good resource for explaining to preschoolers how pumpkins grow. Animals in the ecosystem appear in the illustrations.

Websites

Collards
http://communitygardennews.org/gardenmosaics/pgs/science/english/collards.aspx

This webpage is one of the Science Pages from Garden Mosaics, a program of Cornell University's Department of Natural Resources. Click on "Print Version With Activities" to print a two-page information sheet about collards, with activities to try.

KidsGardening.org
www.kidsgardening.org/node/816

Read about a first-grade class using the garden as the context for their writing and reading in the article, *Linking Literacy and Garden Creatures,* on this website from the National Gardening Association.

Related *Early Years* Blog Posts

The following information is from "Learning About the Butterfly Life Cycle With Local Butterflies," published May 7, 2010:

> *I collected some cabbage white butterfly caterpillars on collard leaves to take to school in a few days and brought them into the house. The next afternoon I noticed a pupa in chrysalis form on the hallway wall, almost 2 meters from the collard leaves. What a journey the caterpillar had made to its pupating place!*

Read more at *http://nstacommunities.org/blog/2010/05/07/learning-about-the-butterfly-lifecycle-with-local-butterflies.*

Additional related blog posts:

"Spring, and Moving on Towards Summer," published April 14, 2011: *http://nstacommunities.org/blog/2011/04/14/spring-and-moving-on-towards-summer.*

"Caterpillars All Around," published May 6, 2009: *http://nstacommunities.org/blog/2009/05/06/caterpillars-all-around.*

"Teach the Lifecycle of a Butterfly and Celebrate 40 Years of Eric Carle's *The Very Hungry Caterpillar,*" published February 17, 2009: *http://nstacommunities.org/blog/2009/02/17/teach-the-lifecycle-of-a-butterfly-and-celebrate-40-years-of-eric-carles-the-very-hungry-caterpillar.*

Chapter 24

Water Works[19]

BEFORE YOU BEGIN THIS ACTIVITY:

Remember that each activity is not a stand-alone science or engineering curriculum. Activities are small steps in a journey of science inquiry, as discussed in Chapter 1. Your students will learn more about this concept and about the nature of science if you use this activity as part of an ongoing exploration of a question, a concept, or a topic being investigated by your class. Ask yourself, "What should come before this and what should come after?" Refer to Table 1.1 (pp. 15–39) to find other activities from *The Early Years* column that address the same concepts.

"You worked hard digging that hole." "You really had to work to push that wagon." "Climbing on the monkey bars is hard work." These are phrases commonly heard on the playground. Children love to try difficult physical tasks that require the expenditure of energy. They know that even if they do not complete a task they will feel the satisfaction of working hard. Tell children, "I've got a job for you," and they eagerly crowd around

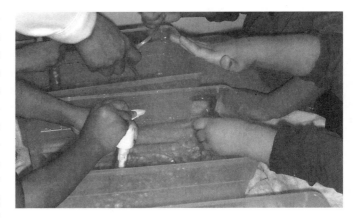

to see what interesting use they will put their bodies to. *Work* is what they will be doing as they move water from one place to another, and the fact that the job has no apparent purpose doesn't seem to bother them. As the teacher, however, you'll know this "job" is giving students valuable practice in working with science tools. *Work* is also a term used to describe the result of *force,* a push or pull acting on an object causing it to move. Making observations about the effects of different strengths or different directions of pushes and pulls on the motion of an object

---

[19] This column entry was originally published in *Science and Children* in July 2007.

is part of the *Next Generation Science Standards (NGSS)* kindergarten performance expectations K-PS2-1 and K-PS2-2 (Motion and Stability: Forces and Interactions): "Plan and conduct an investigation to compare the effects of different strengths or different directions of pushes and pulls on the motion of an object" and "Analyze data to determine if a design solution works as intended to change the speed or direction of an object with a push or a pull" (NGSS Lead States 2013).

Gravity is a pulling force that children are familiar with even if they don't yet know the term—they know that blocks always fall down instead of staying in the air or falling up, and that beverages always go down from the pitcher—not up, not sideways, but down every time. In a pump, such as a liquid soap pump, the liquid inside must go up (from the bottom of the container through the tube to the opening) to come out. This is the opposite of what happens when children pour a liquid but is not likely to be noticed unless a teacher directs attention to it. (An inexpensive way to allow the children to repeatedly use the liquid soap container is to fill it with colored water.)

In the following activity, children will use their own force plus the force of gravity—and simple tools—to move water. Small motor work may not be fun if the child is afraid of failure. The activity introduces the easy-to-use tools first (spoons and scoops, then dental irrigators) and the difficult ones last (eyedroppers, turkey basters, infant nasal aspirators, and pumps). Allow plenty of time for discovery, and repeat the experience so children can thoroughly investigate. Exploring the function of tools and how the shape of an object helps it function as needed to solve a given problem is part of the *NGSS* performance expectation K-2-ETS1 Engineering Design (NGSS Lead States 2013).

Reference

NGSS Lead States. 2013. *Next Generation Science Standards: For states, by states.* Washington, DC: National Academies Press. *www.nextgenscience.org/next-generation-science-standards.*

Activity: How Can We Move Water?

Objectives

- To explore the idea of work as expending energy to move water

- To raise the question of what force(s) move water

- To gain experience using different science tools (spoons, scoops, pipettes, etc.)

Materials

- Two square or rectangular (clear if possible) plastic tubs for each group of two to four students

- Towels

- Water

- Food coloring (optional; makes the water easier to see)

- At least two types of various small tools for moving water: spoons; scoops (laundry powder, coffee, or measuring); various eyedroppers or basters; pipettes; syringe dental irrigators (ask a dentist to donate unused ones); infant nasal aspirators; and toy pumps or pumps from liquid soap bottles (remove the pumps from the bottle and insert them into a piece of clear tubing that is just a few millimeters longer than the intake tube of the pump, to support it).

Procedure

1. Set up plastic tubs side by side with a towel underneath. Fill one about one-quarter full with water.

2. Tell the children that you have a job you want them to do. Ask them to move the water from one tub to another using a spoon.

3. Join in with gusto, and talk about doing work: *I'm lifting the water with the spoon. Do I have to push the spoon or pull the spoon to get it to go down in the other tub?* If no child responds after several times, say, *To put the water into the other tub I have to tip the spoon and the water falls down into the tub. I am not pushing the water down.*

4. Ask, *Will the water ever move up instead of down?* Children may say, "No, it always falls down because of gravity!" depending on prior explanations they've had, yet not understand that gravity is a force. You can put the emphasis on the concept of force, saying, *A force must be pulling the water down.*

5. Wonder aloud how long it will take to move all the water, and solicit ideas about how to make it go faster. If a student suggests lifting one container and pouring all the water into the other, respond positively by saying, *That would work!* and then ask for ideas on how to more quickly move the water without lifting up the containers.

6. One type at a time, provide the additional tools. Many children will need support to work out how to use the pipette-type tools. Ask, *How do these new tools move water? What muscles are you using when you move water with that tool? When does the water go into the pipette—when you squeeze the bulb or when you release it?* Tell students, *You use your force to lift the spoonful of water or to squeeze air out of the pipette. Then water goes in the pipette when the bulb pops back out. What tool do you use your force to push to move the water?* (pump) *What tool do you pull and push to move the water?* (syringe dental irrigator)

7. Ask the students to talk about the tools using one or more of these questions, over time: *What tool works "best" to transfer the water? What tool is easiest to use? What tool is quickest? Do you have to work to lift up the water with that tool? Do you have to work to put the water down with that tool?* Children eagerly try new tools, at first preferring the tool that moves the most water the fastest. Given time, they focus on a particular tool they favor.

8. Putting their tally marks on a teacher-created "Favorite Tool Chart" can help children think about the difference between the tool they liked to use best (and why) and the tool that moved the most water (and why).

9. Sanitize the tools and tubs according to your program's practice and air dry before storing.

Extension

In a follow-up discussion, ask questions about using force to do work. Have students draw their favorite tool and then draw arrows to show where they put their force—lifting, squeezing, pushing, or pulling. As they draw they will talk about the tool and can write or dictate some of their words. Intermediate elementary students may be ready to define the word *gravity*, the attraction between the mass of Earth and the mass of objects on Earth. If a younger student, in preschool to grade 2, uses the word *gravity*, you can say, *We use the word "gravity" to name the force that works to pull things to Earth. Who or what is doing the work to move the water? We use our muscles to lift up the water, and gravity does the work to make it fall into the second container.*

Teacher's Picks

Publications[20]

The Wonder Thing by Libby Hathorn, illustrated by Peter Gouldthorpe (Houghton Mifflin, 1996).

Poetic language and vivid illustrations showing water's beauty make this an excellent book for talking with preschoolers about how important it is to conserve water and not litter or pollute the water supply.

Snowballs by Lois Ehlert (Harcourt Brace, 1995).

This is a cute story for preschoolers about building a snow family that (of course) melts.

Where Do Puddles Go? by Fay Robinson (Rookie Read-About Science; Children's Press, 1995).

This book gives a good explanation of the water cycle, which also mentions water pollution.

It Could Still Be Water by Allan Fowler (Rookie Read-About Science; Children's Press, 1993).

This book explains that water isn't always found in the liquid state; shows other forms such as steam, fog, and ice; and has a nice explanation of parts of the water cycle.

Science With Water by Helen Edom (Usborne Science Activities; Usborne, 1992).

This book has experiments of every kind about the properties of water.

Website

Peep and the Big Wide World
http://peepandthebigwideworld.com/en/educators

This website for preschool educators is in English and Spanish and provides curricula and guidance on teaching science concepts about water, color, plants, ramps, shadows, and sound.

---

[20] These suggestions were provided by Yvonne Fogelman, an educator whose interests included family science and math programs for preschoolers through second grade.

Related *Early Years* Blog Posts

The following information is from "With Water Play Children Gain Experience They Can Record in Writing and Drawing," published November 11, 2009:

> *Playing in water opens many avenues for science explorations, including flow, wetness, phases of matter, volume, and buoyancy. Unexpected results make children think and explore further. For example, children know that fish are supposed to float, so playing with a toy fish that sinks will get children thinking about why. We can let children know that questions are to be shared through our interactions—by listening to their questions, asking open-ended questions, and having the students record their answers or dictate to us to record. Science activities are good platforms for using literacy skills because children often want to tell the story of what they did.*

Read more at *http://nstacommunities.org/blog/2009/11/11/water-play.*

Additional related blog post:

"Discovery Bottles," published September 17, 2009: *http://nstacommunities.org/blog/2009/09/17/discovery-bottles.*

Chapter 25

Counting a Culture of Mealworms[21]

BEFORE YOU BEGIN THIS ACTIVITY:

Remember that each activity is not a stand-alone science or engineering curriculum. Activities are small steps in a journey of science inquiry, as discussed in Chapter 1. Your students will learn more about this concept and about the nature of science if you use this activity as part of an ongoing exploration of a question, a concept, or a topic being investigated by your class. Ask yourself, "What should come before this and what should come after?" Refer to Table 1.1 (pp. 15–39) to find other activities from *The Early Years* column that address the same concepts.

Math is not the only topic that will be discussed when young children are asked to care for and count "mealworms," a type of insect larvae. (Just as caterpillars are the babies of butterflies, these larvae are babies of beetles.) While counting and later sorting the larvae, children will ask questions such as, *What is an insect? Is it alive? Will they bite? Are they eating? How many legs do they have? Where are the mommies and daddies? Why did it change?* They will also begin to understand why scientists count to collect data.

As children collect data, they will closely observe the beetles and discover answers to their questions. The larvae of the darkling beetle (*Tenebrio molitor*) are commonly called "mealworms" even though they are insects with six legs, not worms at all. They will change body forms as they grow to adulthood. First the larvae will get larger, occasionally shedding their exoskeleton, or skin, and growing a new one. Then the larvae will change into pupae, which do not eat and move very

---

[21] This column entry was originally published in *Science and Children* in October 2007.

little. Inside their exoskeletons, pupae are changing into the adult form, beetles with wings.

These insects are fun and easy to care for in the classroom. The larvae or mealworms are slow moving, don't bite, are just weird enough to be interesting, and only need weekly attention. They become livelier as adults, but if they escape they will die and not set up housekeeping in your cupboards. Care is not needed over the weekend, and their container is easy to transport to share with another class when your class is finished.

CAUTION: Biological supply companies and the U.S. Department of Agriculture warn against releasing the beetles into the environment. The entire culture may be frozen overnight to kill any insects that you will no longer care for. Dispose of them according to your school district's policy on organic waste material.

The following activity can take place over two months as the beetles undergo metamorphosis from larvae to adults. As the children care for and count the insects, they will make observations, think about what is alive and what is not, learn what defines an insect, begin to understand life cycles and life spans, and gain big vocabulary words that they love to know and use to feel like "real" scientists.

Although a full understanding about life cycles is not expected until grade 3 in the *Next Generation Science Standards (NGSS)*, observations of animals, including their bodies and their behavior, are part of Life Sciences performance expectations beginning in kindergarten (NGSS Lead States 2013). Children will use three of the *NGSS* science practices (Analyzing and Interpreting Data, Using Mathematics and Computational Thinking, and Engaging in Argument From Evidence) as they count the insects and describe their observations as evidence for their understanding of the changes in an insect life cycle.

Reference

NGSS Lead States. 2013. *Next Generation Science Standards: For states, by states.* Washington, DC: National Academies Press. *www.nextgenscience.org/next-generation-science-standards*.

Activity: Beetle Roundup

Objective

To count, sort, and collect data, while observing the change in body form of an insect as it matures to adulthood (metamorphosis)

Materials

- A culture of mealworms (often sold at pet stores or bait shops, 25 to a small tub)

- Clear container with a lid that allows air flow

- Culture medium (food and bedding): dry, uncooked wheat bran and oatmeal

- Thin slices of apple or potato

- Tray, plastic placemat, or sheet of poster board

- Small bowls for sorting

- Beetle count log sheet, notebook, or poster (see Figure 25.1 for an example)

- Magnifiers

Procedure

1. Find or make a ventilated habitat container for the beetle larvae. Reusing a plastic food container works well.

2. Introduce the children to the larvae by telling them you have baby animals to show them, and then spread the larvae and the medium they came with on a tray to view and gently handle if desired. *CAUTION: Wash hands with soap and water before and after working with insects. Students should not be allowed to eat or drink in the room when working with insects. Using water and detergent, wash any surface areas of desks where insects were placed.*

3. Have the children count the number of live larvae and any dead ones (they become dark and thinner as they dry up).

4. With the children, set up a habitat for the larvae (see resources in "Teacher's Picks" section for additional details) by adding fresh culture medium and several thin, air-dried (to limit mold growth) slices of either apple or potato. *CAUTION: Monitor fruit for mold and remove at first indication of colony growth.*

5. Discuss and display appropriate rules (e.g., "Be gentle") for handling the small animals with the children and have them suggest wording for informative labels (e.g., "These animals are insects and have six legs") for the beetle habitat.

6. Have the children record the date and the number of larvae (live and dead) in a beetle count log before adding the live larvae to the habitat. Columns for recording pupae and adult counts will initially be empty, or you can add the columns later, once children discover larvae that have changed to pupae. More than one entry may be made on each date if additional students count.

7. At least once a week, have students dump the contents of the habitat onto a tray and count the larvae. Several students can work together, gathering the larvae into small bowls, counting, and then adding the bowls together for a total count to be recorded in the log. The children will be excited to see small details using magnifiers. *CAUTION: Always have the children wash their hands with soap and water before and after handling animals.*

8. Continue recording the date and number of animals (larvae at first, later some pupae, and finally adults) for about two months—the minimum time needed for the development into adult beetles. Introduce the term *metamorphosis* when children notice the first pupa. The life cycle is faster in warm classrooms, and it comes to a standstill if the animals are refrigerated. Maintain the culture by adding new food—an apple or potato slice weekly and fresh medium monthly.

9. The results of the beetle count will become more interesting once pupae, and later adult beetles, appear.

Extension

More questions are raised in doing this activity than can be answered solely by counting. Specific to sorting and counting is the question of what to do if another child disputes the count or an individual larva appears to be halfway to its pupa form—does it count as a larva or as a pupa? These are real-life problems that students can solve themselves, thus gaining an appreciation of how scientists make sure they record data accurately. Other questions the children have not yet been able to answer through observation may be researched by reading about mealworms.

After becoming familiar with these small animals, the children can share their knowledge of insect body structure by constructing an insect from precut felt pieces or in drawings.

Teacher's Picks

Publications

Meet the Mealworms by Teena Staller, in *Science and Children* 42 (5): 28–31, 2005.

Finished with sorting and counting? In this journal article you can learn how a fourth-grade teacher and her class began the school year with a 10-day study into animal characteristics and behavior.

A Mealworm's Life by John Himmelman (Nature Upclose; Children's Press, 2001).

By following a mealworm through her life from larva to adult beetle, children come to understand and empathize with the needs of a beetle. The illustrations show the body parts in detail.

Mealworms: Raise Them, Watch Them, See Them Change by Adrienne Mason, illustrated by Angela Vaculik (Kids Can Press, 1998).

This book includes detailed illustrations of the mealworm beetle and interesting experiments to try.

Websites

Q?rius: Insects, Centipedes, Millipedes, Spiders and Relatives
http://qrius.si.edu/browse/type/226

This website from the Smithsonian Institution's National Museum of Natural History allows viewers to search the collection to learn more about insects and other small creatures.

Mealworm Life Cycle
www.enchantedlearning.com/subjects/insects/beetles/mealworm/mealwormlifecycle.shtml

This page on the Enchanted Learning website can be used as a self-check for students who want to include many details in their drawings of the Tenebrio beetle life stages.

Mealworms and Darkling Beetles
www.lawrencehallofscience.org/foss/fossweb/teachers/materials/
plantanimal/tenebriobeetles.html

This page on the FOSSWEB website from the Lawrence Hall of Science includes a "Mealworm Life Cycle" chart and other information. It also has one of the best justifications for having live animals in the classroom: "And they are just nice to have around to remind us that life on Earth takes a seemingly endless variety of forms, and that part of being human is to have compassion and respect for all life."

FIGURE 25.1. Example of beetle count log sheet

Beetle count! How many larvae, pupae, and adults?

Each week count how many there are of each life stage of the beetle. Make a tally mark for each animal you count. Have you observed any changes?

Date	How many larvae? 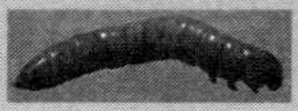	How many pupae? 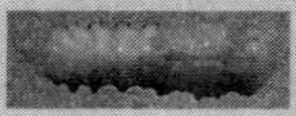	How many adult beetles? 

Related *Early Years* Blog Posts

The following information is from "Children's Drawings Reflect Their Observations—and Their Thoughts," published November 14, 2010:

Early childhood teacher and author Marie Faust Evitt (her book *Thinking BIG, Learning BIG* was published by Gryphon House in 2009) wrote:

> *Saying initial drawings may be more like what they think they should see than what they actually see—is beautiful. I noticed something similar during our recent bean seed sprouting explorations. The first time children drew pictures of their seeds in their science journals, several gave the seeds smiles. Also, they dictated stories about the seeds rather than describing what the seeds actually looked like. Their drawings and observations became more accurate each day as they observed the roots and shoots sprouting and growing. The repeated experiences of recording observations in their journals made the children truly feel like scientists.*

Read more at *http://nstacommunities.org/blog/2010/11/14/childrens-drawings-reflect-their-observations-and-their-thoughts.*

Additional related blog posts:

"East Coast Periodic Cicadas—Prepare Children to Comfortably View These Interesting Insects," published May 30, 2013: *http://nstacommunities.org/blog/2013/05/30/east-coast-periodic-cicadas-prepare-children-to-comfortably-view-these-interesting-insects.*

"Observing, Learning About, Appreciating, and (Maybe) Holding Small Animals Such as Insects," published October 5, 2008: *http://nstacommunities.org/blog/2008/10/05/observing-learning-about-appreciating-and-maybe-holding-small-animals-such-as-insects.*

Chapter 26

Observing With Magnifiers[22]

BEFORE YOU BEGIN THIS ACTIVITY:

Remember that each activity is not a stand-alone science or engineering curriculum. Activities are small steps in a journey of science inquiry, as discussed in Chapter 1. Your students will learn more about this concept and about the nature of science if you use this activity as part of an ongoing exploration of a question, a concept, or a topic being investigated by your class. Ask yourself, "What should come before this and what should come after?" Refer to Table 1.1 (pp. 15–39) to find other activities from The Early Years column that address the same concepts.

Teachers want students to pay attention, to notice patterns in mathematics and language arts, to make observations, and to become astute observers of the natural world. Acknowledging the practice or products of observation encourages this. With a classroom of students it can be difficult to let each child know that you see or hear their discovery. Children's observations may not follow the lesson plan, but they are important to recognize even if they are off topic.

One way to keep the discussion in context without squashing enthusiasm for noticing is to set up a silent signal. Tell the students, *Sometimes there are many people who want to show me their work. I may not be able to talk with each one, but I do hear you and will give you a nod* (or you can use a thumbs-up or "okay" gesture instead) *to let you know.* The silent signal will acknowledge children's observations without extended conversation. Another example of a silent signal acknowledging students is to have children put their hands on top of their heads if they can't see the pictures of a book being read aloud. Instead of saying "I can't see, I can't see!" they silently communicate.

Established silent signals work well to acknowledge those children who hear the distant fire engine in the middle of roll call or see a fly during group instruction at the board and are unable to wait to share their observation. The children have satisfied their need to communicate, and the teacher has encouraged observation without disrupting the direction of the class. Writing or drawing what they observe to be acknowledged later is another way to respect the need for sharing observations without stopping the entire class.

---

[22] This column entry was originally published in *Science and Children* in February 2008.

Inspire your students to become detailed observers by encouraging the use of magnifiers. Magnification can make us see an object with new understanding and is especially helpful for children with low vision. Do you remember the first time you saw the tiny hooklets on the barbules of a barb on a bird feather and understood how the feather can separate and come back together? Rachel Carson said, "Some of nature's most exquisite handiwork is on a miniature scale, as anyone knows who has applied a magnifying glass to a snowflake" (Carson 1956, p. 47).

Magnifying tools point to the interdependence of science (understanding how light travels through matter), engineering (design of the lens and handle), and technology (use of the tool) (NRC 2012). Making observations is part of the science and engineering practices (NGSS Lead States 2013). Students will understand that the material of the magnifier changes what they see, gaining experience that is the foundation for later learning about the properties of light, such as refraction (NGSS Lead States 2013).

Be aware that a child's lack of interest in using a magnifier may be due to preference (other interests prevail), difficulties in near vision or hand-eye coordination, or a short attention span. Using handheld magnifiers requires steady hands and good eye-hand coordination, especially for high-power lenses, so magnifications of 3–6x are most appropriate for young children. The following activity uses interesting objects to give children a reason to learn to use a magnifier. Keep magnifiers in all areas of the classroom to enlarge the area where observations are made—next to the insect collection, hanging near a mirror, and on the library shelf—and keep one in your pocket to use anytime the class is outside.

References

Carson, R. 1956. Help your child to wonder. *Woman's Home Companion,* July. *http:// digitalmedia.fws.gov/cdm/singleitem/collection/document/id/1055/rec/1.*

National Research Council (NRC). 2012. *A framework for K–12 science education: Practices, crosscutting concepts, and core ideas.* Washington, DC: National Academies Press. *www. nap.edu/catalog/13165/a-framework-for-k-12-science-education-practices-crosscutting-concepts.*

NGSS Lead States. 2013. Appendix F—Science and engineering practices in the *Next Generation Science Standards.* In *Next Generation Science Standards: For states, by states.* Washington, DC: National Academies Press. *www.nextgenscience.org/next-generation-science-standards.*

Activity: Exploring Magnifiers

Objectives

- To introduce the hand lens or magnifier and explore its properties

- To become so comfortable using magnifiers that students will use them throughout the year

Materials

- Large columnar clear container, such as a jumbo-size pickle jar, a 2 L bottle with the top cut off, or an acrylic ice bucket

- Water

- Towels or trays

- A variety of magnifiers, including round lenses, bar magnifiers, and flat Fresnel page magnifiers (Most should have a magnification strength of 2–3x, but include a few with 6x power.)

- Interesting objects to look at using magnifiers (e.g., purchased, sterile feathers; grains of salt; leaves; tree bark; hair; cloth; and pictures from magazines and newspapers)

- Two or three small objects (of high interest to children) to drop into water

Procedure

1. To introduce the idea of magnification, fill a large clear columnar jar with water and put it on a table where students can easily view it. Put it in an area of bright light for best viewing.

2. Drop a coin, key, marble, or other desirable object that will sink into the jar, and then reach in to retrieve it. As you reach into the water, the children will see that your hand appears larger in the water.

3. They may say, "Your hand got bigger." Ask them if your hand "grew bigger" or if it "looks bigger." They may be unsure of what happened and will need additional experience with the water-filled jar to understand that what they are seeing has changed, not the size of your hand itself.

4. Provide a setup where the children can repeat this activity themselves with a minimum of spillage. The jar may need to be lowered and put on a

towel or tray. Children will be more interested in additional "doing" than viewing, but while waiting for a turn they will also observe and have time to do a drawing of "what I see."

5. Provide a variety of magnifiers and a set of interesting objects, and have the children practice using magnifiers to examine the details of the objects.

6. Tell the children: *The magnifiers are tools for looking, not for drumming, hitting, digging, poking, or eating.* The words you use will vary depending on the children's age. Some children may need direct instruction on how to get a clear, in-focus view using a magnifier. Naturalists suggest holding the magnifier close to your eye and bringing an object up in front of it until the object is in focus, to give the best view of an object. Until they gain more experience, children may need to hold the magnifier close to the object first, and then bring their head close to the magnifier. This helps them work out how to look at immovable objects and also reduces the number of times children press the magnifier against their eyes. *CAUTION: Sanitize the magnifiers after use.*

7. Tell them, *By moving the magnifier closer to yourself and then farther away, you can find the point where you can see best. Somewhere in between you will see a clear image.* As children develop hand-eye coordination, it is easier for them to view clearly and steadily through a magnifier.

8. After an unstructured period of magnifying the objects, ask open-ended questions about the magnifiers to encourage thoughtful examination of the magnifiers: *What do the magnifiers have in common? How do they look and feel?* Suggest that the children pinch their fingers over the lens, move them from one side to the other, and tell what they feel. Most children describe the magnifier by function, saying, "It makes everything bigger," and "You look through it." Only a few children will say that the lenses are clear and not flat but curved. Suggest additional observation using two magnifiers stacked together. Say, *How does (the object) look now?*

You will know that children are fully comfortable with magnifiers and aware of their purpose when they recognize by themselves that they need to get and use a magnifier to further their exploration or answer their question. Keep magnifiers at hand and it might happen like this: "What are those black dots in the rock?" asks a student. *What could they be?* you reply. "I don't know. I need a magnifier to find out!"

Teacher's Picks

Publications[23]

MoonLight Moon Calendar (Celestial Products, available at *www.MoonCalendar.com*).

This annual poster-size calendar shows the phase of the Moon each night of the year. Looking for the Moon every night is a great opportunity for students and their parents to make observations and study a pattern over time.

A House Is a House for Me by Mary Ann Hoberman, illustrated by Betty Fraser (Puffin Books, 2007).

This book opens up lots of conversation among students and lays the foundation for students building their own analogies by making careful observations about the relationship between animals (and inanimate objects) and their "homes" in the text and illustrations.

Where's Waldo? by Martin Handford (Candlewick Press, 2007).

Use this series when introducing camouflage and the importance of looking carefully to truly see something close at hand. Magnifiers can be used to ease viewing details.

Look-Alikes: The More You Look, the More You See! by Joan Steiner (Little, Brown Books for Young Readers, 2003).

This book invites young readers into "the land of look-alikes, where the more you look, the more you see!" An amazing journey into a world of everyday objects used in ingenious ways, the illustrations compel close observation.

The Jumbo Book of Hidden Pictures (Boyds Mills Press, 1992).

Young readers will eagerly search each picture for the items listed by name.

---

[23] These suggestions were provided by Nancy Tooker, a private school teacher.

Website

Bird Academy: All About Feathers
https://academy.allaboutbirds.org/features/all-about-feathers/#how-feathers-are-built.php

On the "How Feathers are Built" page of this interactive website from the Cornell Lab of Ornithology, students can see an illustration of the details of feathers.

Related *Early Years* Blog Posts

The following information is from "When Young Children Use Magnifiers," published April 21, 2010:

> *The children have used magnifiers many times before but they are always intrigued, and maybe even a little surprised, when things "get bigger." One child asked why the magnifier made things bigger. I had him feel the shape of the lens to feel how the plastic curved and told him the curve bends the path of light so the image looks bigger. I don't expect him to understand all of that, but I held my finger under a magnifier and asked, "Does my finger look bigger or did it really get bigger?" The children said it just looked bigger but felt my finger to check! They used an additional sense (touch) to verify their observation.*

Read more at *http://nstacommunities.org/blog/2010/04/21/when-young-children-use-magnifiers*.

Additional related blog post:

"Magnifiers," published November 16, 2009: *http://nstacommunities.org/blog/2009/11/16/magnifiers*.

Chapter 27

Objects in Motion[24]

BEFORE YOU BEGIN THIS ACTIVITY:

Remember that each activity is not a stand-alone science or engineering curriculum. Activities are small steps in a journey of science inquiry, as discussed in Chapter 1. Your students will learn more about this concept and about the nature of science if you use this activity as part of an ongoing exploration of a question, a concept, or a topic being investigated by your class. Ask yourself, "What should come before this and what should come after?" Refer to Table 1.1 (pp. 15–39) to find other activities from The Early Years column that address the same concepts.

Objects in motion attract children. Because they remain interesting all year long, small spinning tops are a good choice for long-term residency on the science table, along with other tools: magnifiers, tweezers, and pipettes. Although tops are not tools for manipulating objects, they are tools for exploring laws of motion.

Tops are available in many forms. Even 2-year-olds can spin the kind of non-electric top that is a flat disk with a shiny pattern, sometimes called a laser or Mylar spinning top (see examples of this type of top at *www.ustoy.com/lazer-tops-2-25*). Putting different types together on a tray invites comparison—how easy is it to spin the big top compared with the small one? You can also have children compare flat versus tall, wood versus plastic, and solid versus hollow tops. With continued experience using tops of different sizes and shapes, children learn that the top will spin faster and longer if they push faster as they spin it or spin it on a smooth table rather than on a cloth and that the top will remain upright while it spins fastest but begins to tilt as it slows to a stop. Open-ended questions can help children express what they have learned and to think further: *What do you notice about the top? Tell me what you see. Why did that top keep spinning? How are those tops the same?*

A marble race is another high-interest tool for exploring motion. Students can build "tracks" using cardboard tubes taped at various angles to a vertical surface such as a wall or large box. Young children also delight in making a ramp to roll

---

[24] This column entry was originally published in *Science and Children* in March 2008.

toy cars. Children understand that the small cars zooming down the ramp are models that represent actual cars.

The following activity helps children explore the motion of bodies riding in a vehicle and safely demonstrates the answer to their question, "Why do I need a seat belt?" Children will enjoy moving the cup around, even if all they "see" is a cup rather than understanding that it represents a car. They will understand that each time they suddenly stop the cup, the marble will roll out unless it is taped in, even if they do not yet understand that the same forces apply to passengers in a car.

Exploring motion addresses the *Next Generation Science Standards (NGSS)* kindergarten performance expectation K-PS2, Motion and Stability: Forces and Interactions (NGSS Lead States 2013), about investigating pushes and pulls on the motion of objects. Understanding about the nature of science also develops as children work as scientists to search for cause-and-effect relationships to explain natural events (e.g., "The marble rolled because I pushed it"), which is one of the crosscutting concepts in the *NGSS* Appendix G.

Reference

NGSS Lead States. 2013. *Next Generation Science Standards: For states, by states.* Washington, DC: National Academies Press. *www.nextgenscience.org/next-generation-science-standards.*

Activity: Crash Dummy Fun!

Objective
To provide experience using models to explore motion and force

Materials

- Clear plastic cups

- Marbles

- Tape

Procedure

1. Gather the children (entire class or small group) into a circle. Hold up a clear plastic cup and a marble and say, *This is my pretend car. I'm going to go for a drive, and this marble is me.*

2. Holding the cup on its side with the open end forward, place the marble inside the clear plastic cup and push the car-cup forward slowly across the floor, and then slowly stop it, saying, *Brrrmmm, brmm, brrrmmmm. Stop sign.* The marble should stay in the cup.

3. Then go out for another "drive," this time stopping suddenly to avoid another "car." Say, *Brmmm, brmmm. Uh-oh! That other car moved too close in front of me. I have to stop suddenly so it doesn't bump my car!* (Make the sound of a tire screech.) Stop quickly enough that the marble rolls out of the cup.

4. Ask the children to tell you what happened and what happened to "you." Children will often say, "The marble fell out of the cup." Remind them that the cup and marble represent a car and driver by saying, *Yes, I fell out of my car.*

5. Ask for suggestions on what you could do to stay safely in your car if the situation ever comes up again. Many children say "Close the door!" meaning to close the open end of the cup. Put your hand over the opening and move the car back and forth vigorously to show that the marble still moves inside the cup just as a person inside a car would.

6. Refine the question by saying, *In real life—not pretend—how can we make sure that we don't keep going when our car stops? What holds us safely in the car and in our seats if we have to stop the car very suddenly?* Children usually say, "Put on your seat belt!"

7. Put a piece of tape across the marble still inside the cup. Move and stop the cup quickly to show how the "seat belt" keeps the marble (driver) in the car.

8. Have all the children use a cup and marble as a car-and-driver model, first with and then without a seat belt.

Children do not hesitate to crash their car-cups into each other. Their smiles and quick glances at the teacher show that they know they are representing a danger-ous action, safe to do only with the cup-as-car models. After "falling out of the car" a few times, most ask for a piece of tape to use as a seat belt. For a follow-up discussion, you could ask, *When else have you seen objects continue to move once their carrier has stopped?* Some examples include groceries on a checkout conveyor belt, people on a sled that is stopped by a snowdrift, paint on a spin art turntable, and items on a lunch tray that is pushed too quickly along the line.

Teacher's Picks

Publications

Tires, Spokes, and Sprockets: A Book About Wheels and Axles by Michael Dahl, illustrated by Denise Shea (Amazing Science: Simple Machines; Picture Window Books, 2006); and *Forces Make Things Move* by Kimberly Brubaker Bradley, illustrated by Paul Meisel (Let's-Read-and-Find-Out Science, Stage 2; HarperCollins, 2005).

These two nonfiction books about motion can be used as a refresher for teachers or a further exploration for students.

ABCDrive! A Car Trip Alphabet by Naomi Howland (Clarion Books, 1994); *Alphabeep: A Zipping, Zooming ABC* by Debora Pearson, illustrated by Edward Miller (Holiday House, 2003); *Are We There Yet, Daddy?* by Virginia Walters, illustrated by S. D. Schindler (Viking Juvenile, 1999); and *A Drive in The Country* by Michael J. Rosen, illustrated by Marc Burckhardt (Candlewick Press, 2007).

These four books—ABC collections of vehicles and stories of families riding in cars—have illustrations showing adults and children wearing seat belts. A search game can be made for illustrations where a child is riding in the front seat or other hazardous behavior.

The Ball Bounced by Nancy Tafuri (Greenwillow Books, 1989) and *Wheel Away!* by Dayle Ann Dodds (HarperCollins, 1991).

These two books for the youngest children show a ball and a wheel in motion, staying in motion until acted on by an outside force.

Websites

Booster Seat Flyer
http://depts.washington.edu/booster/educational_materials.html

Order these free one-page informative flyers (updated in 2007) from the Washington State Booster Seat Coalition to raise awareness about booster seat basics. The flyers are available in many languages.

Car Seat Types
www.safercar.gov/parents/CarSeats/Car-Seat-Types.htm

This section of the National Highway Traffic Safety Administration website lists the different car seat restraints appropriate at different ages.

Related *Early Years* Blog Posts

The following information is from "Children and Motion," published October 2, 2011:

> *What is in motion in your classroom—in addition to children? Spinning tops are one of the materials I keep available all year long because they can be an independent or collaborative activity, children's ability to spin them increases as they grow, and spinning tops is an exploration in physical science, teaching children about pushes and pulls. Spinning tops can be part of learning mathematics. Young children can sort tops by size, weight, and shape, then record which top needs the biggest twist-push to begin spinning and which top spins the longest.*

Read more at *http://nstacommunities.org/blog/2011/10/02/children-and-motion.*

Additional related blog posts:

"Early Education in Engineering and Design," published October 12, 2013: *http://nstacommunities.org/blog/2013/10/12/early-education-in-engineering-and-design.*

"Learning About Motion and Appropriate Restraints," published February 9, 2009: *http://nstacommunities.org/blog/2009/02/09/learning-about-motion-and-appropriate-restraints.*

Chapter 28

An Invertebrate Garden[25]

BEFORE YOU BEGIN THIS ACTIVITY:

Remember that each activity is not a stand-alone science or engineering curriculum. Activities are small steps in a journey of science inquiry, as discussed in Chapter 1. Your students will learn more about this concept and about the nature of science if you use this activity as part of an ongoing exploration of a question, a concept, or a topic being investigated by your class. Ask yourself, "What should come before this and what should come after?" Refer to Table 1.1 (pp. 15–39) to find other activities from The Early Years column that address the same concepts.

The early summer is the time to lay the groundwork for a lesson that will begin in September: finding invertebrates outside and learning how they connect to that location and the plants they were found near. Small invertebrate animals found in a school yard may be representatives of three animal categories or phyla—Annelida (segmented worms), Arthropoda (spiders, insects, crustaceans, and more), and Mollusca (snails and slugs). Whether or not they are desirable depends on the viewpoint of your community. For farmers and gardeners, slugs and snails may be serious pests that will limit the amount of harvest, but for a child they represent a world to be explored.

For teachers, invertebrates are tools for broadening students' understanding about animals and the connections between animals and habitats or plants; invertebrates are also an engaging topic to write or read about. Studying the small animals that inhabit a garden

---

[25] This column entry was originally published in *Science and Children* in July 2008.

supports the *Next Generation Science Standards (NGSS)* kindergarten performance expectation K-LS1-1: Use observations to describe patterns of what plants and animals (including humans) need to survive (NGSS Lead States 2013). In *A Framework for K–12 Science Education* (*Framework*; NRC 2012), the disciplinary core idea in Life Sciences LS1.C for grades K–2 asks, "How do organisms obtain and use the matter and energy they need to live and grow?" This is part of the foundation for *NGSS* performance expectations.

Use the last few weeks of school or the first few weeks of summer to establish sites to attract invertebrates, so they will be available to study when September arrives. If you know your area and use the "Look, don't touch" rule for students when observing insects in the garden, invertebrate gardens can be a great learning opportunity for students. Attract invertebrates with plantings and temporary structures designed to provide the food, shelter, and breeding needs of particular terrestrial invertebrates that are safe to be around. Use the resources listed at the end of the activity description to find out about invertebrates commonly used in classroom studies, as well as those that are not used, because they bite, sting, or are poisonous.

CAUTION: In any exploration of living organisms, there are safety considerations. When using structures, do not use complex structures that could provide places for possibly dangerous invertebrates to hide in. For example, a pile of old boards is a good place to find isopods but may also harbor brown recluse spiders, a poisonous species. Just one board, laid flat on the ground, is enough to shelter isopods, millipedes, and slugs yet allow easy sighting of any unsafe residents such as centipedes or poisonous spiders. Similarly, many flowering plants will attract both butterflies and stinging insects such as bees or wasps. Know the contents of the site's soil before planting there, as well as any poisonous species that may be in your area. Be aware of garden surroundings when working there with students, and make sure all students wash hands thoroughly after working in the garden.

Enlist the support of any staff that may be in the building over the summer—summer school teachers, custodians, and administrative staff. Explain your goal of having an invertebrate-friendly garden and the need for the vegetation pile to remain undisturbed over the summer and to have access to water. Then be sure to let them know you appreciate their contribution to the project with a student-written note, a mention in the school newsletter, or even a plate of homemade cookies!

References

National Research Council (NRC). 2012. *A framework for K–12 science education: Practices, crosscutting concepts, and core ideas.* Washington, DC: National Academies Press. *www.nap.edu/catalog/13165/a-framework-for-k-12-science-education-practices-crosscutting-concepts.*

NGSS Lead States. 2013. *Next Generation Science Standards: For states, by states.* Washington, DC: National Academies Press. *www.nextgenscience.org/next-generation-science-standards.*

Activity: Attracting Invertebrates

Objective

To plant food source plants and create conducive environments to attract invertebrates for study in the fall

Materials

- Plants, including parsley, fennel, milkweed, goldenrod, violet, zinnia, aster, daisy, pansy, lettuce, and collard. (*Note:* Mature plants have a better chance of surviving untended.)

- Potting soil and container or garden plot

- Leaf mulch or peat moss

- Vegetation pile of plant matter such as leaf mulch, grass clippings, a straw bale, and vegetation from your yard or kitchen—any fruit or vegetable scraps but no meat, dairy, or oils

- Untreated wood to frame the compost

- Large flat rocks or concrete chunks

- Gallon jugs with a pinhole in bottom

- Digging tools, such as a shovel, trowels, soup spoons, and garden or spading fork (Note: A soup spoon with a sturdy handle is ideal for digging a small hole; these spoons can be purchased inexpensively at thrift stores.)

- Gloves

Procedure

1. With your class (if school is still in session), set up a garden bed or container garden in a sunny spot and a place for a compost pile in a shady spot.

2. An adult can loosen the soil with a spading fork or shovel before the children smash any dirt clods and add organic matter if needed—leaf mulch or peat moss—with soup spoons or trowels.

3. Plant seedlings or mature plants for larval and adult food sources. Select plants that will attract the species of invertebrates you are interested in studying, telling your class, *We're going to see what invertebrate animals might be here when school begins again. Invertebrates we often see around here are ____.*

(Name a few that are common to your location, such as bees, roly-polies, butterflies, slugs, or snails.)

4. Set up piles of yard waste (collected leaves, dead plants, and grass cuttings) or a straw bale in a loose frame of boards laid end to end, and water well. Put the large flat rocks or concrete chunks next to the pile. Use your judgment in having children help handle the boards and rocks— avoid boards with splinters and help children carefully place rocks.

5. Have the students wash hands after preparing the garden.

6. If possible, return weekly to add to the vegetation pile. Large pieces such as cantaloupe rinds and thick broccoli stalks can provide moisture for some weeks.

7. Water the plants and pile as needed. The decomposing matter should be damp underneath but not soggy. Fill the pierced gallon jugs with water and place next to the most water-needy plants to provide a second day's watering.

8. When school begins, look for milkweed bugs and monarch butterfly caterpillars on the milkweed; other caterpillars on the collard, violet, parsley, and fennel; ladybug larvae on the daisies; butterflies on the zinnia, goldenrod, and asters; and slugs and snails on the lettuce and pansies. Wearing gloves, lift up the board, rocks, and vegetation pile to look for isopods, slugs, snails, earthworms, crickets, and millipedes. Early in the day can be a good time to observe the invertebrates in the garden.

It is hard to find one book or website that has information about all the invertebrates that students observe. Supplement your invertebrate study with a library search and/or invite a guest gardener or entomologist to help identify the invertebrates observed.

Resources

Caras, R., and S. Foster. 1994. *A field guide to venomous animals and poisonous plants: North America north of Mexico.* Peterson Field Guides. New York: Houghton Mifflin.

Eidietis, L., S. Gray, L. Riggs, B. West, and M. Coffman. 2007. Goodbye critter jitters. *Science and Children* (45) 1: 37–41.

Full Option Science System (FOSS), Materials management: Plant & animal care. *www. fossweb.com/plant-animal-care.*

Xerces Society for Invertebrate Conservation. Identification guides. [Various titles.] *www. xerces.org/publications/identification-guides.*

Teacher's Picks[26]

Publications

Bugs! Bugs! Bugs! by Bob Barner (Chronicle Books, 1999).

Even the youngest bug enthusiast will find these colorful collage illustrations and the rhyming text engaging. Eight common "bugs" are featured. The final pages list details and life-size renderings for students fascinated by the facts.

From Caterpillar to Butterfly by Deborah Heiligman, illustrated by Bari Weissman (HarperCollins, 1996).

Butterflies may be the most easily appreciated of all insects. This book follows the growth of a painted lady butterfly raised in a classroom, from caterpillar to adult flying out the window. A brief final section introduces other common butterfly species and lists butterfly centers that can be visited.

Life on a Little-Known Planet by Howard Ensign Evans (Lyons Press, 1993).

This is a good book for teachers that highlights the wonderfully strange habits of insects, richly described by an entomologist and naturalist.

Handbook of Nature Study by Anna Botsford Comstock (Comstock, 1986).

Found a praying mantis, a walking stick, or firefly and want to know what it eats, where it lives, and how it defends itself? This is a great book for teachers to demonstrate to students what it means to "look it up." It also has terrific ideas for investigations.

Websites

Kids' Inquiry of Diverse Species
www.biokids.umich.edu

The BioKids website is a partnership of the University of Michigan School of Education, the University of Michigan Museum of Zoology, and the Detroit Public Schools. Teachers can use the "Field Guides: Invertebrate ID Guide" to narrow the

---

[26] These suggestions were provided by Fred Arnold, a science resource teacher.

field and identify the collected invertebrates, and students can view photos in the "Critter Catalog" to find a match and learn a little about the animal.

What's That Bug?
www.whatsthatbug.com

Found a cool bug but no one seems to know what it is? This website was set up to identify insects from submitted photographs. Chances are good that you can find your interesting insect discovery in their enormous library of identified insects, just by browsing the site. Not there? You can send in a picture which, in time, may be identified.

Related *Early Years* Blog Posts

The following information is from "Ecosystems Outside the School Door," published March 6, 2014:

> *Since it is now March and in my area we just had our 10th snow day, I am dreaming of planting seeds rather than actually planting them. I asked the children what we should plant in the raised bed school garden and they offered very general ideas: "flowers," "carrots," and "peas" (from the older children who remember them from last spring). These are great ideas and we can do them all—first peas and later carrots and flowers, all in small amounts. Like many schools, we don't have a windowsill with full sunlight where we can grow healthy seedlings inside. We do sprout seeds in plastic bags, in clear containers, and on sponges, to see the delicate roots and sprouts grow. But growing a plant to maturity must wait until we can plant in the ground, currently under a blanket of snow.*

Read more at *http://nstacommunities.org/blog/2014/03/06/ecosystems-outside-the-school-door*.

Additional related blog posts:

"Update on the Success of Using Local Butterflies," published June 16, 2010: *http://nstacommunities.org/blog/2010/06/16/update-on-the-success-of-using-local-butterflies*.

"Invertebrate Diversity Tally," published October 16, 2012: *http://nstacommunities.org/blog/2012/10/16/invertebrate-diversity-tally*.

"Invertebrates in the Classroom," published December 8, 2008: *http://nstacommunities.org/blog/2008/12/08/invertebrates-in-the-classroom*.

Chapter 29

Air Is Not Nothing[27]

BEFORE YOU BEGIN THIS ACTIVITY:

Remember that each activity is not a stand-alone science or engineering curriculum. Activities are small steps in a journey of science inquiry, as discussed in Chapter 1. Your students will learn more about this concept and about the nature of science if you use this activity as part of an ongoing exploration of a question, a concept, or a topic being investigated by your class. Ask yourself, "What should come before this and what should come after?" Refer to Table 1.1 (pp. 15–39) to find other activities from The Early Years column that address the same concepts.

"If I can't see it, it doesn't exist." This is a "given" for many children, which can make "air" a tricky topic to address with this age group. Children usually begin to understand that a material called air is all around us after age 3, but they don't grasp that air is matter until age 5, or even older. They may learn that air is a gas but have difficulty naming the material that fills a soap bubble or explaining how a balloon expands, and they don't understand where a gas released by opening a soda or mixing baking soda and vinegar comes from or where it goes. Children don't often recognize that plants need air to grow.

Yet, amid these ideas, early childhood is rich with opportunities for students to experience a range of behaviors of gases, even if children can't name the gases or explain their behaviors.

How do objects move through air? This question is answered by children as they explore many fun activities in early childhood—from dropping sippy cups to waving flags to throwing balls. These activities may become more complex with developmental

[27] This column entry was originally published in *Science and Children* in December 2008.

level: 5-year-olds delight in running across the playground with a scarf streaming behind them like a superhero's cape, 6-year-olds enjoy the back-and-forth folding to create a personal handheld fan, 7-year-olds fold paper into shapes for flight, and 8-year-olds want to measure and graph how far a single breath can blow a feather from a straw.

Exploring the motion of air and introducing the vocabulary of states of matter (e.g., *gas*) support children's learning about the properties of materials, including the gas they are so familiar with. Understanding about molecules and changes in matter won't be well developed until grade 5 or later. However, the *Next Generation Science Standards* performance expectations in kindergarten (K-PS2-1: "Plan and conduct an investigation to compare the effects of different strengths or different directions of pushes and pulls on the motion of an object") and grade 1 (1-PS4-1: "Plan and conduct investigations to provide evidence that vibrating materials can make sound and that sound can make materials vibrate") include investigations that involve the presence of air (NGSS Lead States 2013). By providing experiences and time to reflect on them and discuss ideas, we can support young children's understanding of matter.

Let children feel the material of air by holding a sheet of cardboard and fanning it up and down. Later, presenting children with a range of everyday materials to sort according to the state of matter—including ambiguous items such as a bag of salt (pours but is not wet), a bag of air, and shaving cream foam (wet and can be shaped to fill different containers but doesn't flow)—is a way of encouraging participation and thinking about matter (Varelas et al. 2008). Young children won't yet grasp these concepts, but it is important to provide them with early explorations so they can build a body of experience to recall as they move to deeper understandings in the future.

References

NGSS Lead States. 2013. *Next Generation Science Standards: For states, by states.* Washington, DC: National Academies Press. *www.nextgenscience.org/next-generation-science-standards.*

Varelas, M., C. C. Pappas, J. M. Kane, A. Arsenault, J. Hankes, and B. M. Cowan. 2008. Urban primary-grade children think and talk science: Curricular and instructional practices that nurture participation and argumentation. *Science Education* 92 (1): 65–69.

Activity: Moving Air

Objective
To experience air's mass and the force it can exert on objects

Materials

- Balloon (for the teacher to demonstrate with) *CAUTION: Balloons are a choking and allergy hazard for young children. Teachers should also be aware of latex allergies.*

- Drinking straw for each child

- Feather for each child (*Note:* Using one color of feathers will help focus the attention on what happens rather than on favorite colors.) *CAUTION: Check for allergies to feathers, and never use nonsterile wild feathers; purchase sterile feathers at a craft store.*

- Cotton balls

- 1-liter soda bottles (*Note:* Narrow mouth works best. Several bottles are needed because they become flattened with use.)

Procedure

1. Blow up a balloon with exaggerated big breaths. Ask the students, *What is happening to the balloon?* If needed, prompt with additional questions, *Why is it getting bigger? What am I putting into the balloon?* They may say, "You're blowing your breath into it," but not many will use the word *air,* and only a few will say "a gas." Tell the class that only the teacher is allowed to put his or her mouth on the balloon.

2. Let go of the balloon, allowing the force of the balloon contracting and the air flowing out to move the balloon erratically around the room, for fun and to demonstrate how a gas can move an object.

3. Give each child a drinking straw with a feather partially inserted into one end and tell them to use their breath to blow the feather out, up to the ceiling or sky. The children will experiment with different-size breaths or may ask for a different-size feather.

4. Ask the children to talk about what is making the feathers move and why sometimes the feathers go farther than others.

5. Have the children insert a cotton ball into but not through the opening of a 1-liter soda bottle. Tell them to use both hands to quickly squeeze the body of the bottle to force the air out, popping the cotton ball up and out. Young children show that this is an unexpected event by their surprise—they didn't know that air can push. Children with enough

hand strength, generally age 5 and older, will enjoy squeezing the bottle to eject the cotton ball, but it is also instructional for younger children who may inadvertently suck the ball into the bottle and then must spend a long time figuring out how to get it out. If they can't shake it out, they may try using their fingers or use a pencil, or they may turn the bottle upside down to bring the cotton ball closer to the opening. Before a child becomes frustrated enough to give up, suggest that they ask an experienced classmate or yourself for ideas to try. If the bottle is held upside down with the cotton ball close to the opening, a quick forceful squeeze usually blows the cotton ball out.

6. Whenever you have an opportunity while investigating air, discuss the children's ideas about air and the recorded results to find out what they think about air. "It's this stuff … you can't see it, but it's really there," sums up a beginning understanding of gas. Some children may suggest ideas for further exploration.

Many scientists began important work when they wondered about everyday experiences that everyone else always took for granted. Teachers and families do important work when we encourage children to ask questions and look for answers—especially when our expectation is that they will have additional exploration to do before grasping the concepts.

Teacher's Picks[28]

Publications

How the Ladies Stopped the Wind by Bruce McMillan, illustrated by Gunnella (HMH Books for Young Readers, 2007).

Creative women in a windy Icelandic village join forces with chickens and cows to solve the problem of the disappearing trees. Brightly colored illustrations add to the fun.

---

[28] These suggestions were provided by Marie Faust Evitt, a preschool teacher and author of the book *Thinking BIG, Learning BIG: Connecting Science, Math, Literacy, and Language in Early Childhood* (Gryphon House, 2009). The book's website has a useful "Links/Resources" tab: *http://thinkingbiglearningbig.com.*

Can You See the Wind? by Allan Fowler (Rookie Read-About Science; Children's Press, 1999) and *Feel the Wind* by Arthur Dorros (Let's-Read-and-Find-Out Science, Stage 2; HarperTrophy, 1990).

These two nonfiction books explain the basics of wind in language children can understand and relate to. Use these books to help children learn what causes wind, how people can use its power, and how it affects weather.

Blow Away Soon by Betsy James, illustrated by Anna Vojtech (Putnam Juvenile, 1995).

Help children face their fear of the wind by reading the story of Sophie and her Nana, who build a "blow-away-soon" from desert treasures near their home in the Southwest. This book could lead into an experiment testing which items blow in the wind and which stay put.

Websites

Weather Wiz Kids
www.weatherwizkids.com

Meteorologist Crystal Wicker created this educational website for teachers and students. It presents background information on all types of weather, including wind, thunderstorms, and tornadoes. The Kid's Zone section of the website includes jokes, games, folklore, flashcards, and links to other weather-related websites.

Wind
www.need.org/wind

This website from the National Energy Education Development Project presents a wealth of information and lesson plans (for elementary grades through grade 12) about wind and wind power as a renewable energy source. See the *Primary Energy Infobook* (pp. 67–70) and the *Wonders of Wind* Teacher and Student Guides for background information.

Related *Early Years* Blog Posts

The following information is from "What Shape Is Your Bubble Wand? Children and Making Choices," published May 2, 2009:

> *Children like to have choices (as do I). Choosing their favorite color pipe cleaner (also known as "fuzzy sticks"), marker color, place in line, type*

of seed to plant, or which center in which to begin their day can be so important for young children that they are willing to overcome shyness or difficulty with language to voice their choice. Being encouraged to choose and plan helps children develop thinking, and talking about what they might do in the future is important for science because part of being able to make predictions is to think about what has not yet happened.

Read more at *http://nstacommunities.org/blog/2009/05/02/what-shape-is-your-bubble-wand*.

Additional related blog posts:

"Air Is Matter," published December 11, 2008: *http://nstacommunities.org/blog/2008/12/11/air-is-matter.*

"Observing Closely—Bubbles!" published December 10, 2010: *http://nstacommunities.org/blog/2010/10/10/observing-closely%E2%80%94bubbles.*

Chapter 30

Bring on Spring: Planting Peas[29]

BEFORE YOU BEGIN THIS ACTIVITY:

Remember that each activity is not a stand-alone science or engineering curriculum. Activities are small steps in a journey of science inquiry, as discussed in Chapter 1. Your students will learn more about this concept and about the nature of science if you use this activity as part of an ongoing exploration of a question, a concept, or a topic being investigated by your class. Ask yourself, "What should come before this and what should come after?" Refer to Table 1.1 (pp. 15–39) to find other activities from The Early Years column that address the same concepts.

As the month of February begins, it's hard to believe that the year is turning to spring! Season's change may at first seem imperceptible unless we measure its arrival somehow. As a way to measure spring's arrival, young children can plant certain seeds outside (as soon as the soil can be worked), and older students can record the sunrise time on a chart. Students may not be able to understand why the Sun rises earlier every week as the year moves to spring, but they can see that there is a change and also see the change in a sprouting seed.

In U.S. Department of Agriculture (USDA) agricultural zones 6 and 7 (see "Resource" section at the end of the activity description), peas can be planted on or around President's Day in mid-February (February 16 in 2015); it helps to have prepared the garden bed ahead of time before winter. With this early start, the plants have time to grow, bloom, and set fruit, and the pods have time to

Sprouting pea seeds

---

[29] This column entry was originally published in *Science and Children* in February 2009.

grow to eating size before mid-June when many schools close for the summer. Growing peas helps children who grow food at home make a connection between school and home by sharing their experience, and helps those who have never seen a field of crops make a connection between soil, weather, and food production by growing a trial crop. Observing plants to describe what growing plants need to survive is part of the *Next Generation Science Standards* kindergarten performance expectation K-LS1-1, "Use observations to describe patterns of what plants and animals (including humans) need to survive" (NGSS Lead States 2013).

For those in other plant hardiness zones, refer to the seed packet and plant either earlier or later than Presidents' Day, or try planting another early crop, such as radishes. If you do not have outdoor space for planting, smaller varieties of peas may do well in containers.

Call seed companies and request a catalog or two to keep in your classroom. An internet search for "seed catalogs" produces a long list of companies. Use the catalogs to show your students that foods of all kinds are grown from seeds, and use the pictures to plan a garden—real or pretend. If you order some seeds, use the photos to make labels for your plants. Even fourth- and fifth-grade students may not know where their food comes from before it gets to the store (Rubenstein et al. 2006).

Planting seeds indoors as well as outside allows the students to see the details of a sprouting seed and compare the growth of plants with drastically different amounts of sunlight and nutrition. Planting a seed in dirt or a clear container such as a plastic bag is an experience that students can refer to in later years when studying plant cells and how a root grows. The nutrients in the pea seed's cotyledon (seed leaves) will nourish the plant for a time, but without additional nutrients (minerals provided by soil and food made by the plant using sunlight) pea plants growing indoors will become weak. Teachers can try keeping them alive long enough to compare with the seeds students plant outside by providing sunlight and a weak solution of plant fertilizer.

References

NGSS Lead States. 2013. *Next Generation Science Standards: For states, by states.* Washington, DC: National Academies Press. *www.nextgenscience.org/next-generation-science-standards.*

Rubenstein, H., A. C. Barton, P. Koch, and I. R. Contento. 2006. From garden to table: Rural or urban, two effective strategies that teach students about where food really comes from. *Science and Children* 43 (6): 30–33.

Activity: Planting Peas and Observing Growth

Objective

To notice and participate in the arrival of spring by planting seeds and observing them sprout, grow, and produce fruit

Materials

- Pea seeds (*Note:* The best seeds are the edible-podded varieties with the shortest time to maturity and needing little support.) *CAUTION: Avoid the poisonous ornamental sweet peas. Rinse the seeds in advance to remove any pesticides.*

- Magnifier

- Resealable plastic bags

- Paper towels

- Permanent marker

- Stapler

- Tape

- Water

- Ruler (measuring centimeters) or nonstandard measuring tool such as cubes or links

- Outside garden area or container of potting soil

- Digging tools (e.g., soup spoons)

Procedure

1. Talk with the class about any signs of spring they have noticed—earlier sunrise, longer days, occasionally warmer weather, sighting of certain

animals, sprouting leaves and buds of spring flowering bulbs, and the flowering of bushes such as forsythia, Japanese quince, and pussy willow.

2. Ask students when it will be time to plant outside—how do we know? Accept all answers, and then refer to the USDA Plant Hardiness Zone Map (see "Resource" section) and say, *Farmers and scientists have been measuring the temperature all over the world every day for many, many years. This map shows zones, or areas, of similar temperature to show where different plants can survive and grow.* Older children can find their location on the zone map.

3. Have the children examine the pea seeds with a magnifier. Some may express surprise that the peas they eat for dinner are also seeds, showing how little they know about how food is grown.

4. Have students soak the seeds in water for several hours, then place them on top of a damp paper towel (folded into fourths) inside a resealable plastic bag—one for the class or one for each child. Using a permanent marker, draw a line across the width of the bag about 10 cm from the bottom, and have the children staple along the line. The staples should be 3–8 mm apart (to prevent the seeds from falling to the bottom but not stapled so tightly that the roots can't get through to the water that will be at the bottom of the bag). Add three to five seeds and enough water to make a small reservoir at the bottom (about 2 cm deep), then tape to a wall to maintain the bag in an upright position.

5. Ask the children, *What will happen to the seed as it begins to grow?* Children's comments reveal their prior knowledge, ranging from a shrug to "It's going to get bigger!" to "It needs dirt," to the more detailed description, "The root pushes out and grows down."

6. Daily, have students measure and record any growth of root and sprout with a centimeter ruler or relative measurement, such as "smaller than a pencil point," and draw what they see.

7. Nutrients stored in the seed's cotyledon will nourish the plant for a time, but without additional nutrients (minerals provided by soil or fertilizer), and food (made by the plant using sunlight) the pea plants will become weak in about a month. A discussion about the needs of living things may inspire students to try to find out what the pea plants need and are not getting.

8. If feasible, at the same time peas are placed in bags indoors, have the children also plant peas outside in a garden area or large pot, about 1 inch deep (for the children, one pinky finger deep). Soup spoons are good digging tools for young children. On any day that it is not snowing or

raining, have the students water the garden. Compare these plants with the seeds sprouted inside.

Children's interest level will vary as they wait for the seed to sprout, as the plant grows, and as the flowers and then fruits appear. Record when the children first notice flowers and fruit. Celebrate your food production success with a snack of peas, augmented with some purchased from a farmer or grocery store if necessary.

Resource

U.S. National Arboretum. 2012. USDA plant hardiness zone map. *www.usna.usda.gov/ Hardzone.*

Teacher's Picks[30]

Publications

Botany on Your Plate by Katharine D. Barrett (National Gardening Association, 2008).

Although developed for early elementary grades, this standards-based life science curriculum provides a lot of good ideas for the prekindergarten classroom, too. It inspires children to explore plants parts that we eat, weaving together science and nutrition.

Gardening Wizardry for Kids by L. Patricia Kite, illustrated by Yvette Santiago Banek (Barron's Educational Series, 1995).

This kid-friendly book that includes colorful histories and folklore of common fruits, vegetables, and herbs, along with indoor growing projects and engaging investigations for teachers to use with their students.

Oliver's Vegetables by Vivian French, illustrated by Alison Bartlett (Hodder Children's Books, 1995).

Where do French fries comes from? Your class might have their own ideas. Oliver discovers the answer to this question on a trip to his grandparents' house.

---

[30] These suggestions were provided by Sarah Pounders, a youth education specialist at the National Gardening Association.

Tops and Bottoms by Janet Stevens (Harcourt, 1995).

An entertaining story of an energetic hare outsmarting a lazy bear by knowing the different edible parts of plants, this book teaches children about these parts as you read aloud.

Websites

Harvest of the Month
www.harvestofthemonth.com

This website from the California Department of Public Health includes an "Educator's Corner" page with activities, resources, recipes, and more.

KidsGardening
www.kidsgardening.org

This website from the National Gardening Association provides activity and lesson ideas, horticultural background materials, information about grant and award opportunities, and a monthly e-newsletter, *Kids Garden News.*

Poisonous
http://gardening.ces.ncsu.edu/plants-2/poisonous

It is always important to keep safety in mind. This website from North Carolina State University and North Carolina A&T State University provides an extensive list of poisonous plants.

School Garden Wizard
www.schoolgardenwizard.org

This website provides a detailed guide to beginning a school garden program.

My First Garden: Teacher's Guide
http://extension.illinois.edu/firstgarden/tg

This website from the University of Illinois Extension offers in-depth gardening information at an introductory level for new gardeners.

Related *Early Years* Blog Posts

The following information is from "Planting Peas—Who Will Help Students Record the Growth?" published February 7, 2009:

> *I'm wondering what crops your class grows—peas? collards? cilantro? zinnias? marigolds? In USDA plant hardiness temperature zone 7a, late February is the earliest time I can plant cool-weather crops. But this year with temperatures varying from 21°F to 51°F in the same week, I'm never sure if the children will be able to plant on days when I'm teaching at their school. Garden results have been mixed. One year we had enough pea pods for every child to try one. They loved the crunch! Another year every pea shoot was eaten to a stub, perhaps by voles. Then there was the year there was so little rain and not enough human waterers, resulting in stunted pea plants that produced only a few pods.*

Read more at *http://nstacommunities.org/blog/2009/02/07/planting-peas*.

Additional related blog post:

"Spring Activities and Books to Go With Them," published May 12, 2012: *http://nstacommunities.org/blog/2012/05/12/spring-activities-and-books-to-go-with-them*.

Chapter 31

Does Light Go Through It?[31]

BEFORE YOU BEGIN THIS ACTIVITY:

Remember that each activity is not a stand-alone science or engineering curriculum. Activities are small steps in a journey of science inquiry, as discussed in Chapter 1. Your students will learn more about this concept and about the nature of science if you use this activity as part of an ongoing exploration of a question, a concept, or a topic being investigated by your class. Ask yourself, "What should come before this and what should come after?" Refer to Table 1.1 (pp. 15–39) to find other activities from The Early Years column that address the same concepts.

Words give us the power to describe our world and how we experience it. Any time we classify something, we give it a name to distinguish it from all others of its kind. Like the buttons on a kitchen blender that say "mix" in five or six different ways, there is more than one word to describe if or how light travels through a material. *Opaque, translucent,* and *transparent* are appropriate words for young children because they can be distinguished by observation and without measuring.

For this activity, children can describe a material as *opaque* if no light passes through it, *translucent* if some light passes through, and *transparent* if they cannot tell if any light is blocked, that is, it appears that all the light passes through the material. New vocabulary words such as *opaque* only become important if we need to use them. A student might ask, "What is a word for when you can't see through something but it lets some light through?" Once a new word is introduced to the class, a teacher can use it frequently at

---

[31] This column entry was originally published in *Science and Children* in March 2009.

first to help students remember it. *The translucent bathroom window lets light in, but I can't see what is happening outside; I can't see through the box to see what is inside because it is opaque; or I can see what the fish is doing because its aquarium is transparent* are ways teachers might use the words once they are introduced.

Shadow puppetry is a fun activity where the property of opacity is important. Put on a shadow puppet show with your class (see "Teacher's Picks" section later in this chapter). Have students try it first with a translucent screen such as a white plastic trash bag and then with opaque and transparent screens. Which type of material is needed to best display the shadows? Where can the audience view the shadows?

Working with a variety of materials and a light source will prepare students to consider in later years the nature of light and how it can be blocked (because the light is either absorbed or reflected, or both) or partially blocked by some objects. Children may need some practice controlling a flashlight before being asked to focus on systematically testing the opacity of the materials. In the following activity, children test the opacity of various materials and learn about light. This investigation helps children build understanding of physical science concepts in the *Next Generation Science Standards* grade 1 performance expectation 1-PS4-3 ("Plan and conduct an investigation to determine the effect of placing objects made with different materials in the path of a beam of light") and the grade 2 performance expectation 2-PS1-1 ("Plan and conduct an investigation to describe and classify different kinds of materials by their observable properties") (NGSS Lead States 2013).

Reference

NGSS Lead States. 2013. *Next Generation Science Standards: For states, by states.* Washington, DC: National Academies Press. *www.nextgenscience.org/next-generation-science-standards.*

Activity: Playing With Light

Objective
To explore how light passes through, or is "blocked" by, transparent, translucent, and opaque materials

Materials

- Source of light accessible to the students: a light table, a flashlight, or a sunny window

- Squares (each at least 12 cm square) of the following materials: aluminum foil, cardboard, or other opaque material; acetate, clear plastic, or other transparent material; wax paper, thin tightly woven cloth, tissue paper, or other translucent material

- Paper for a Prediction Chart (what we think will happen) and a Results Chart (what we saw), enough for individual or group recording

Procedure

1. Provide the materials at an open-ended center for children to explore for a week or more. *CAUTION: Teach children how to use flashlights safely. Tell them that they should never shine lights in their eyes or anyone else's eyes.*

2. In a small or large group, hold a flashlight and a square of material to cover the lighted end, and ask students to predict if they will be able to see the light shining when the material is covering the light source. (Some children may mistakenly think you mean to cover the entire flashlight, so be prepared for understanding to develop as you present additional materials.)

3. Record, or have the students record, each answer on a Prediction Chart (what we think will happen) in words, with a check in the appropriate column ("It will block the light"; "Some light will shine through"; "All the light will shine through"), or with symbols such as a dark circle, small Sun, and large Sun. Most children predict, "The light will go through it (the clear plastic)," and many children predict that the cloth will block the light, especially if it is a dark color.

4. Accept and record all on-topic answers. If conflicting predictions are given, begin a brief discussion of how scientists don't always agree about the ideas they are exploring and say, *Perhaps through our work we'll come to an agreement about what we think does happen.*

5. Turn off the lights and shine the flashlight through each material in turn. With each material, ask the class to tell you what they see: no light, a little light, or a lot of light. Shine a light through your hand by cupping it over a flashlight.

6. Have the students record their observations on an individual or group Results Chart (what we saw). Young children can tape a small piece of each material onto the appropriate space marked with words and a symbol, such as "no light" and a dark circle.

7. Have students compare the materials that are classified in the different groups and say how the materials are the same and how they are different. Many children will initially say that the opaque materials do not have any holes in them to let the light through. Side-by-side comparison with a transparent film or the translucent tissue paper, for example, shows them that there is another property that operates here, allowing light to pass through or not.

Children's ideas and wonderings about light can reveal what they think light is. Some may say, "The light gets blocked by the aluminum foil but only some gets caught in the cloth," indicating they made accurate observations. (Young children won't realize that some of the light is also blocked when a flashlight is covered by a colored film.) Ask them if they see the light that is caught, to encourage further thinking about the nature of light.

Extension

Test the opacity of a variety of natural materials, such as leaves, slices of wood, water (liquid and frozen), slices of rock, and feathers. Ask the children to find something in the classroom that light will pass through, or *something that does not block, or take away any of the light.* Children will notice that some materials let more light through than others. Further activities can address whether materials absorb or reflect light.

Teacher's Picks

Publications

The Dark, Dark Night by M. Christina Butler, illustrated by Jane Chapman (Good Books, 2008).

This slightly scary story relates how the animals are at first fooled by the size of shadows created by lantern light. Shadows can appear larger than the object blocking the light. Read it aloud and ask your class to apply what they learn to their explorations in shadow puppetry.

"Science 101: Why Are There So Many Different Models of Light?" by William C. Robertson, in *Science and Children* 46 (2): 56–60, 2008.

Read this article about how models can help explain the behavior of light, and do the activities to prepare for teaching about light.

Worms, Shadows, and Whirlpools: Science in the Early Childhood Classroom by Karen Worth and Sharon Grollman (Heinemann, 2003).

In Chapter 4, the section "In the Classroom With Ms. Tomas" (pp. 130–141) details how the class's shadow exploration develops.

Guess Whose Shadow? by Stephen R. Swinburne (Boyds Mills Press, 1999).

As the title suggests, use this book as a guessing game.

Shadows and Reflections by Tana Hoban (Greenwillow Books, 1990).

Have your students search the photographs for familiar shadows and identify opaque, translucent, or transparent objects.

Website

Playing With Shadows
https://artsedge.kennedy-center.org/multimedia/series/AEMicrosites/ playing-with-shadows.aspx

This lesson plan on shadow puppetry from ArtsEdge, an education program of The Kennedy Center, can be adapted for your grade level. You can use a white plastic trash bag taped between two chairs for the shadow screen and use a flashlight lantern as the light source if an overhead projector is not available.

Related *Early Years* Blog Posts

The following information is from "Two-Year-Olds Explore Transparent, Translucent, and Opaque Materials," published January 29, 2010:

> *Science activities with 2-year-olds usually do not last very long, but sometimes the children surprise me. One group of four children spent about 15 minutes exploring a set of cardboard tubes with ends covered with clear plastic wrap, wax paper, or a double layer of black plastic (black construction paper would also work). We looked through the tubes*

and talked about what we saw—could we see through them? Then I put out small flashlights. Exploration took off! The children tried each tube, comparing how much they could see through the material and how much light from the flashlight came through.

Read more at *http://nstacommunities.org/blog/2010/01/29/two-year-olds-explore-transparent-translucent-and-opaque-materials.*

Additional related blog post:

"Involving Families in Early Childhood Science Education," published February 4, 2012: *http://nstacommunities.org/blog/2012/02/04/involving-families-in-early-childhood-science-education.*

Chapter 32

Hear That?[32]

BEFORE YOU BEGIN THIS ACTIVITY:

Remember that each activity is not a stand-alone science or engineering curriculum. Activities are small steps in a journey of science inquiry, as discussed in Chapter 1. Your students will learn more about this concept and about the nature of science if you use this activity as part of an ongoing exploration of a question, a concept, or a topic being investigated by your class. Ask yourself, "What should come before this and what should come after?" Refer to Table 1.1 (pp. 15–39) to find other activities from The Early Years *column that address the same concepts.*

Like breathing, the ability to hear sound is often taken for granted unless it becomes impaired. Children may not wonder about how sound is generated or detected until introduced to an inquiry activity about sound. Begin by playing a "Stop and Listen" game to help students become more aware of sounds in the classroom or playground environment. Give the class a few minutes to listen, then say, *Stop!* and have the children list and group the sounds they hear. What were the quietest and loudest sounds? Did everyone hear all the sounds? Why or why not? Young children often assume that what they experience is also what others experience. Could all sources of sound be identified? Grouping the sounds into the categories of "Sounds of Nature" (e.g., bird song) and "Human-Made Sounds" (e.g., traffic) may encourage students to listen for small sounds such as those made by grass or leaves blown by the wind.

Play the game outside and ask, *What can you hear in the weather—sunshine, clouds, wind or wind moving something else, rain falling or raindrops hitting something, snow falling, or ice melting?* Simulate raindrops falling on a surface by spraying water from a mister over a sensory table filled with water, and ask the children to describe what they hear. Have they ever heard the sound of rain striking a puddle or of air bubbles popping as rainwater filters down into the ground when the rain stops? How about the sound of ice creaking as it melts and moves? Listening for small sounds is useful later, when you can tell a noisy class to be as quiet as a grain of sand blowing across the ground or an earthworm entering its burrow.

---

[32] This column entry was originally published in *Science and Children* in April 2009.

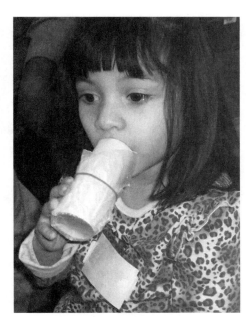

Exploring how to make sounds lays a foundation for children's understanding, much later, of wave properties. The *Next Generation Science Standards* in kindergarten focus on the effects of pushes and pulls, and the grade 1 physical sciences performance expectation 1-PS4-1 states, "Plan and conduct investigations to provide evidence that vibrating materials can make sound and that sound can make materials vibrate" (NGSS Lead States 2013). Ask your class, *How can we make sound?* (Be prepared for voices calling, shouting, and whistling; clapping hands; and drumming feet!) Because air is a somewhat mysterious substance to young children, the idea of waves going through the air is difficult to understand. For that reason, combining *feeling* sound while hearing it is a good way to introduce the topic.

Older people sometimes try to impress upon younger generations that you can "make do" and have a lot of fun with the materials around you—such as a comb and piece of tissue paper. By wrapping the tissue around the comb and holding it lightly in your lips as you hum, you can add an entertaining "buzz" sound to your voice, also felt with your lips. Young children's enthusiasm may lead to a soaking wet tissue, so a more practical setup for early childhood teachers who want to teach about sound and vibration is to use "old-time" knowledge to make a cardboard tube kazoo. A *kazoo* is a tool that adds a buzz to your voice as you hum or sing into it. Unlike a whistle, you don't blow—your breath does not go through the kazoo but just fills the open space. The vibration of the kazoo is obvious as sound is produced—a revelation for young children and some older ones as well.

Reference

NGSS Lead States. 2013. *Next Generation Science Standards: For states, by states.* Washington, DC: National Academies Press. *www.nextgenscience.org/next-generation-science-standards.*

Activity: Make a Kazoo

Objective
To feel vibrations that create sound

Materials

Provide one for each child:

- Cardboard tube (reuse empty paper towel tubes)

- Square of wax paper (about 10 cm square)

- Rubber band to hold the wax paper on the tube

Procedure

1. Make a model ahead of time (see step 5). Demonstrate how sounds made with a kazoo have a characteristic buzzing quality.

2. Have each child use one finger to feel the vibration of the wax paper on your kazoo as you first remain silent, and then say their name into the kazoo. They will probably say, "It tickles!" followed by, "I want to do it!"

3. Pass out one cardboard tube, one square of wax paper, and one rubber band to every child.

4. The students can decorate the tubes and label them with their names.

5. Have students wrap the wax paper over one open end of the cardboard tube—taut like a drum head—and secure with the rubber band. Depending on their amount of experience, children as young as age 4 will be able to stretch and twist the rubber band over the tube, but many will need help. Learning to twist the rubber band to secure it over the tube a second time takes practice.

6. Tell the children not to blow but to hum or talk into the tube with their mouths very close to, but not pressed up against, the tube. An easy sound to start with is an exaggerated train whistle, "toot, tooooot." (They will probably blow air into the tube regardless, discovering that they can blow off the wax paper cover, but will eventually get back to making sound.)

7. Sing a familiar song together using the kazoos to enhance the sound.

8. Assess student understanding by asking, *What do you feel? What happens to the kazoo as you make sounds?* When the wax paper is not stretched tight like a drum head, the vibration is reduced. Brainstorm with the class why that happens by asking them how to fix a "broken" kazoo. The youngest children will have no idea, but further experience with rubber bands stretched over an open box, drum heads, or guitar strings will support their understanding that the tautness of the material affects the sound produced.

Some of the young rappers in your class may enjoy teaching others how to make a punctuated, buzzing sound and other mouth percussion elements. Some classes might want to create a body percussion song and then write a simple musical score by using invented symbols for each hand clap, foot stomp, or thigh pat. For more information, see the Kentucky Educational Television lesson plan on body percussion in the "Resource" section.

Resource

Kentucky Educational Television. 2015. Arts toolkit. Lesson plan: Body percussion. *www. ket.org/artstoolkit/music/lessonplan/135.htm.*

Teacher's Picks

Publications

After reading these books, ask your class to come up with a list of sounds that are particular to their own environment.

Marsh Music by Marianne Berkes, illustrated by Robert Noreika (Millbrook Press, 2000).

Frogs of the marsh are realistically illustrated in watercolor (how appropriate!) as they create a symphony of sound, singing and playing spiderweb string instruments—a nice blend of realism and fantasy. The text spans from preschool to music conductor language, including terms such as *woink-woink* and *adagio* (musical terms are defined in a glossary).

Rain by Peter Spier (Yearling, 1997).

Young children can easily show what one must do to make a sound with a triangle or box guitar, but they have a difficult time putting these ideas into words, even when it is obvious that the teacher, with closed eyes, cannot see their gestures. This book has many gestures and no words. Ask your students to tell you the words to fit the pictures, and then make the sounds that go with the pictures.

Listen to the Desert/Oye al Desierto by Pat Mora, illustrated by Francisco X. Mora (Clarion Books, 1994).

The English and Spanish text and earth-tone illustrations poetically present the sounds made in a desert environment.

The Indoor Noisy Book by Margaret Wise Brown, illustrated by Leonard Weisgard (Harper & Row, 1942).

As you read this classic story, have your class listen with the dog, Muffin, to the sounds heard inside a home, and then have them update the list with sounds from their homes.

Websites

Sound Activities
www.wildmusic.org/en/aboutsound/soundactivities

Try some activities and online games from the website of the traveling exhibit "Wild Music," from the Science Museum of Minnesota and the Music Research Institute, University of North Carolina at Greensboro.

Introduction to What Sound Is
www.fearofphysics.com/Sound/dist.html

This section of the website FearOfPhysics.com provides an introduction for teachers on how sound is generated, with graphics for deeper understanding.

DSOKids Listen by Instrument
www.dsokids.com/listen/by-instrument/.aspx

Ask your students to think about how the sound is made as they listen to clips of instruments of the orchestra, one at a time, at this website from the Dallas Symphony Orchestra. Then play a guessing game and try to distinguish between instruments by sound alone. Invite older students to demonstrate playing an instrument.

Related *Early Years* Blog Posts

The following information comes from "Exploring Sound and Music as Part of Science Learning," published December 15, 2010:

> *I like to use a musical triangle to focus children's attention on the tiny movement that generates sound. They touch the still triangle and then remove their hand. I strike the triangle to make the sound and they touch it again (and shiver or giggle as the vibration tickles their fingers). Next we pass the triangle and striker around the circle and one child holds the triangle while the other uses the striker. This way no one person can monopolize the equipment and we all get to feel the vibrations again.*

> *Sometimes in passing the triangle a child will hold the side of the instrument rather than the string, and the resulting "thunk" (rather than a musical note) opens a discussion of how to change sounds.*

Read more at *http://nstacommunities.org/blog/2010/12/15/4992.*

Additional related blog post:

"Feeling Vibrations," published April 5, 2009: *http://nstacommunities.org/blog/2009/04/05/feeling-vibrations.*

Chapter 33

Adding Up the Rain[33]

BEFORE YOU BEGIN THIS ACTIVITY:

Remember that each activity is not a stand-alone science or engineering curriculum. Activities are small steps in a journey of science inquiry, as discussed in Chapter 1. Your students will learn more about this concept and about the nature of science if you use this activity as part of an ongoing exploration of a question, a concept, or a topic being investigated by your class. Ask yourself, "What should come before this and what should come after?" Refer to Table 1.1 (pp. 15–39) to find other activities from The Early Years column that address the same concepts.

Having your class measure and record the amount of precipitation that falls daily is a job young children can do as part of learning about measurement and weather. Although the youngest scientists may not understand the reason why a rain gauge must have straight sides or how a measurement scale is adjusted to correspond with the area of the opening, they will understand that they are catching and measuring the amount of rain that fell in one day. Precipitation and how it comes about may become more interesting, and long-term patterns may be noticed, if it is discussed daily (even briefly). The data can be recorded with drawings (pictures of clouds, raindrops, and snowflakes, and coloring in the water level on a rain gauge drawing) or by writing numbers.

Young children may need repeated measuring tasks before understanding the meaning of the units (Smith, Sterling, and Moyer-Packenham 2006). Doing a comparison activity can support children's understanding about measuring *inches* of rain rather than volume. Display two clear cylindrical containers of different diameters but the same height (1- and 2-liter soda bottles can be cut to the same height). Tell the children that during a rain event, the number of water drops that fall in one spot is about the same as in the spot next to it, so the containers will fill to the same height. Set the cylinders in a water table, or outside, and sprinkle "rain" from a watering can or jug with holes poked in it. Have the children compare, measure, and record the heights of the water column in each container. This activity promotes discussion

---

[33] This column entry was originally published in *Science and Children* in July 2009.

and will prepare children to understand the data they collect as they measure rainfall outside their classroom. Assign the task of measuring the inches of water in the gauge as one of the class jobs, and you will not have to remember the task yourself.

Measuring precipitation could be part of the activities your class does to meet the *Next Generation Science Standards (NGSS)* kindergarten performance standard K-ESS2 Earth's Systems (NGSS Lead States 2013), as children use and share observations of local weather conditions to describe patterns over time. Read about Patterns, one of the crosscutting concepts, in Appendix G of the *NGSS* to learn how concepts can bridge across disciplines and unite core ideas in science and engineering. Read Appendix H in the *NGSS* to learn more about the K–12 progression in understanding the nature of scientific knowledge.

References

NGSS Lead States. 2013. *Next Generation Science Standards: For states, by states.* Washington, DC: National Academies Press. *www.nextgenscience.org/next-generation-science-standards*.

Smith, L. A., D. R. Sterling, and P. S. Moyer-Packenham. 2006. Activities that really measure up. *Science and Children* 44 (2): 30–33.

Activity: Measuring Up

Objective
To measure and collect data about precipitation (rain, hail, and snow)

Materials

- Rain gauge (*Note:* A funnel opening reduces splash-out, and a flat bottom allows accurate measurement.)

- Drawing or writing materials to record measurements

- Internet access (optional)

Teacher Preparation

Learn more about precipitation data collection by viewing the Community Collaborative Rain, Hail and Snow Network (CoCoRaHS) training slideshow. (Optional: Sign up to participate in the citizen-scientist CoCoRaHS network. CoCoRaHS is a nonprofit, community-based network of volunteers, including young children and their teachers, working together to measure and map all forms of precipitation—rain, hail, and snow. The data are used by the National Weather Service, city utilities [water supply, water conservation, storm water], ranchers and farmers, and many others.) The CoCoRaHS online training makes it easy for teachers to understand how a rain gauge can be used to add to a national database. Students will learn about science as a human endeavor and how scientists collaborate to gather knowledge as their data are added on the website.

Procedure

1. Read one of the "Teacher's Picks" resources about rain to your class and begin a discussion about precipitation experiences, what rain is, where rain comes from, and how people use rain. Introduce the word *precipitation* so children will understand that rain and snow are grouped together as water that falls from the sky. Make or purchase a rain gauge.

2. In the classroom, practice measuring amounts of water in the gauge. Students will find that the gauge must be kept level to accurately read the amount.

3. Set up the rain gauge in a location far away from tall vegetation and buildings, and elevate it to improve "gauge catch."

4. Every morning at the same time, have the class measure the amount of precipitation in the gauge. Be as consistent as possible in the location and the period of time the gauge is in place to most accurately monitor precipitation. (Optional: If you have signed up to participate in the CoCoRaHS network, log in to the CoCoRaHS website and enter your data.)

5. Display the measured amount of precipitation on a yearlong calendar to make it easier to see long-term patterns (e.g., Does it really only rain on weekends?). One way to display the data is to post it along with the number of days in school thus far—every day adding a number and the data—eventually circling the room. The data can be seen all year long, and seasonal patterns may be noticed.

Children will eagerly take their turn measuring the amount of precipitation in the gauge and recording the data with a drawing or a number. After exploring rain data in their own area, students may be ready to learn about water distribution around the world (Morgan and Ansberry 2015). When there is no rain for months, add another measuring activity such as measuring plant growth.

Reference

Morgan, E., and K. Ansberry. 2015. Teaching through trade books: All the water in the world. *Science and Children* 52 (9): 16–21.

Resource

Community Collaborative Rain, Hail and Snow Network. (CoCoRaHS). *www.cocorahs.org*.

Teacher's Picks[34]

Publications

Come On, Rain! by Karen Hesse, illustrated by Jon J. Muth (Scholastic, 1999) and *Monsoon* by Uma Krishnaswami, illustrated by Jamel Akib (Farrar, Straus and Giroux, 2003).

Compare and contrast the experiences of children in the two stories—anticipating storms and celebrating the arrival of a downpour in different settings and cultures. How does Tessie's experience of dancing with her friends and their mothers in the streets compare with a young girl's reaction to the start of monsoon season in India?

Rain by Peter Spier (Yearling, 1997).

This book illustrates, but doesn't describe with text, the many ways children joyfully experience rain. Ask your students to tell you the words to fit the pictures, and then make the sounds that go with the pictures.

Rain Song by Lezlie Evans, illustrated by Cynthia Jabar (Houghton Mifflin, 1995).

This story celebrates the sounds of a thunderstorm. After reading about how two children use drums to imitate the beating rain, explore with students how to create their own stormy noises.

---

[34] These suggestions were provided by Marie Faust Evitt, a preschool teacher and author of the book *Thinking BIG, Learning BIG: Connecting Science, Math, Literacy, and Language in Early Childhood* (Gryphon House, 2009). The book's website has a useful "Links/Resources" tab: *http://thinkingbiglearningbig.com*.

Where Do Puddles Go? by Fay Robinson (Rookie Read-About-Science; Children's Press, 1995).

Help make an abstract concept—the water cycle—real for young children using the color photos and simple text describing falling rain and evaporating puddles. This book could lead to observations and experiments testing how quickly water evaporates on pavement at school.

Websites

The Water Cycle—USGS Water Science School
http://ga.water.usgs.gov/edu/watercycle.html

This U.S. Geological Survey (USGS) website contains a wealth of information for teachers and students about every aspect of the water cycle, including an interactive diagram available in numerous languages. Use links to calculate how many baths you can get from a rainstorm (*http://water.usgs.gov/edu/qa-home-baths.html*), and learn about the real shape of raindrops (*http://water.usgs.gov/edu/raindropshape.html*).

The Weather Dude
www.wxdude.com

Meteorologist Nick Walker presents information about all types of weather in a readable form. Follow the links for information about precipitation, clouds, the water cycle, and weather forecasting.

Related *Early Years* Blog Posts

The following information is from "Evaporation—Children Need to Know the Word and Concept," published April 11, 2010:

> *There are many good books that touch on the subject of evaporation. Young children are just beginning to build an understanding of what "matter" is, and the concept of evaporation is one they experience rather than fully understand. Although we don't need to dwell on it, it is part of our vocabulary in early childhood classrooms as we help children understand that water (and all matter) does not "disappear" or "go away" but always exists, although in different forms.*

Read more at *http://nstacommunities.org/blog/2010/04/11/evaporation-children-need-to-know-the-word-and-concept.*

Additional related blog posts:

"Observing Weather Events," published January 3, 2013: *http://nstacommunities.org/blog/2013/01/03/observing-weather-events.*

"Data Collection in Early Childhood," published January 12, 2011: *http://nstacommunities.org/blog/2011/01/12/data-collection-in-early-childhood.*

Chapter 34

Planting Before Winter[35]

BEFORE YOU BEGIN THIS ACTIVITY:

Remember that each activity is not a stand-alone science or engineering curriculum. Activities are small steps in a journey of science inquiry, as discussed in Chapter 1. Your students will learn more about this concept and about the nature of science if you use this activity as part of an ongoing exploration of a question, a concept, or a topic being investigated by your class. Ask yourself, "What should come before this and what should come after?" Refer to Table 1.1 (pp. 15–39) to find other activities from The Early Years column that address the same concepts.

There is much to discover when children plant flower bulbs in the fall. Soil can be compacted and hard to dig in some places and loose in others; its color may differ from spot to spot; and children may encounter small animals such as centipedes, worms, and insect grubs as well as rocks and trash in the soil. Encourage questions by taking a notepad outdoors and writing down what the children ask and speculate, to be researched and discussed later. Mark on your calendar when to look for sprouting leaves so the class can begin to measure the first growth (late winter, early spring). Visit the site often with the children to check for growth, and wonder aloud, *When will the bulbs grow?*

Planting flower bulbs is a wonderful activity for many reasons: learning about the life cycle of a plant bulb teaches children about seasonal changes and the environmental needs of plants, and children can observe and measure plant growth over time and see the results of their work in the spring. Conversation about where to plant can build awareness of how the outdoor space is used by other people and animals, where the Sun shines on the ground, and where the rain falls; it also draws attention to soil as a resource. Planting flower bulbs is also an inexpensive way to have a beautiful garden the children will be proud of. Observing plant growth relates to the *Next Generation Science Standards* kindergarten performance expectation K-LS1, From Molecules to Organisms: Structures and Processes (NGSS Lead States 2013).

---

[35] This column entry was originally published in *Science and Children* in September 2009.

Even when given a reason to dig in the soil, inexperienced children may still hesitate because they are so often told to not mess up their clothes or someone's garden bed. If your class does not regularly dig in sandboxes or outside, students may need practice before they can plant. Practice digging in a sandbox of damp sand, planting seeds in potting soil in paper cups, and digging with spoons in playdough—all indoor experiences that can prepare a child to dig outside. Prepare the children to dig with confidence by sending a note home for families asking the children to dress in old clothes for the day and reassuring all that everyone will wash hands afterward. *CAUTION: In addition to making sure that children wash their hands after digging in soil, train them to keep their hands and plant parts out of their mouths and to dig toward themselves to keep from throwing dirt. As another important safety measure, you must carefully choose bulbs that are not toxic (see "Resources" section later in this chapter).*

Children need to be outside where the complex, ever-changing environment stimulates their brain. Playing outside and planting outside lead children to make discoveries and have experiences that indoor play with potting soil cannot replicate. Understanding is built by having the actual experiences and observing flower buds growing and changing color.

Reference

NGSS Lead States. 2013. *Next Generation Science Standards: For states, by states.* Washington, DC: National Academies Press. *www.nextgenscience.org/next-generation-science-standards.*

Activity: Some Like It Cold

Objectives

- To learn about the needs of flowering bulbs
- To observe seasonal changes in plants

Materials

- Outdoor area for planting—anywhere that will get at least four hours of direct sunlight and can be watered. *CAUTION: Select an area free of natural hazards (e.g., poison ivy) or artificial hazards (e.g., pesticides).*

- Organic materials to amend soil if necessary

- Spring-flowering bulbs of nontoxic species, such as *Camassia spp.* (also called camas, quamash, and wild hyacinth). (See the "Resources" section for information about bulbs and a list of safe and unsafe plants. Note that many bulbs need a period of cold weather to grow flowers—check with your local extension service or garden center. *CAUTION: The best safety practices rely on the classroom teacher knowing the students and the materials and being watchful during science activities.*)

- Books about planting flowers, such as *From Bulb to Daffodil* by Ellen Weiss

- Drawing materials

- One large shovel or pitchfork for teacher use

- Small trowels or large spoons

Teacher Preparation

Prepare the planting area by digging up shovelfuls of soil and turning them over to loosen. Amend the soil with organic material if needed.

Procedure

1. Read a book to the class about planting flowers, including those grown from bulbs, such as *From Bulb to Daffodil* (Weiss 2007). Discuss how plants have needs all year round. They sprout and grow best in certain seasons, and we plant them at the time when the weather best suits their needs.

2. Show the children a bulb, cautioning them that it is not food, and pass it around for close observation. Ask students to *keep it whole* so they don't pick at the bulb, pulling off any loose "skin." Tell students the bulb is not a seed but is like a seed in that it will grow into a particular kind of plant when it has everything it needs to grow well. Many children will chime in with "Rain!" or "Water!" or "Sunshine!" Tell them that in addition to sunlight, water, air, and soil, some plants need a period of cold to grow best and to produce flowers. That is why you will be planting these bulbs before winter, so they can be exposed to cold temperatures.

3. Cut one bulb in half so the children can see that the inside is layered like an onion and the plant leaves are not yet visible. Have students draw the bulb while talking about any prior bulb-planting experiences. Ask, *When will the bulb grow roots and leaves?*

4. Read aloud the directions for planting the bulbs (from the vendor or those you find online) and rehearse the procedure, allowing time for questions. Rehearsing makes it easier for novice gardeners.

5. Take the children outside to plant, providing large spoons or small trowels to a few at a time.

6. While some students are planting, the rest of the class can list all the living organisms and nonliving things they see or listen to another book about planting flower bulbs.

7. Encourage the children to examine and describe the soil while digging and to notice the texture and any components. Have them draw a picture of what the bulb looks like in the ground.

8. Wash hands after planting.

Spring's arrival seems a long time away from fall for young children. Help them remember their planting activity by posting some drawings in the classroom and returning to the bulb garden occasionally throughout the winter to check for sprouting bulb leaves (these will not be seen until early spring). If you record air temperature, you can remind the class that the bulbs are getting the cold period they need to flower, and when it rains say, *The bulbs under the soil are getting the water they need to grow roots and be ready to sprout in the spring.* Children can record their observations on all the visits, even if they just draw soil. After the leaves sprout, drawings get more exciting and observations can include daily measurements of growing leaves and flower stems.

Reference

Weiss, E. 2007. *From bulb to daffodil.* New York: Children's Press.

Resources

Bale, S. S. 1995. *Spring, summer, and fall bulbs.* University of Kentucky, College of Agriculture, Cooperative Extension Service. *www2.ca.uky.edu/agc/pubs/ho/ho80/ho80.pdf.*

University of California, Division of Agriculture and Natural Resources. 2012. Safe and poisonous garden plants. *http://ucanr.edu/sites/poisonous_safe_plants.*

Teacher's Picks

Plants by Sue Barraclough (Investigate!; Heinemann Library, 2008).

With a question-and-answer format, this book asks readers to answer before turning the page.

From Bulb to Daffodil by Ellen Weiss (Children's Press, 2007).

This excellent book includes photos of plant structure details and introduces some vocabulary, so it can be useful for English-language learners as well as early readers. I wish the book did not use the word "sleeping" instead of "leaf senescence" to describe the leaf and flower die-back, because young children can learn, and like to use, big words that are more precise. "Senescence"—I clap and say it to help myself remember: sen-es-ence, (sənes'-əns).

The Life Cycle of a Flower by Molly Aloian and Bobbie Kalman (Crabtree, 2004).

Photographs reveal the details of flower structures and plant parts, and the text describes seed production and other ways plants reproduce.

Boxes for Katje by Candace Fleming, illustrated by Stacey Dressen-McQueen (Farrar, Straus and Giroux, 2003).

This story about tough times in post–World War II Holland is based on the experiences of the author's mother, who sent boxes to a family in Holland.

A Year in the City by Kathy Henderson, illustrated by Paul Howard (McGraw-Hill, 1996).

Some seasonal changes are specific to urban environments: snow gray with car exhaust, Chinese New Year parades, and city park garden blooms.

When This Box is Full by Patricia Lillie, illustrated by Donald Crews (Greenwillow Books, 1993).

Children love to guess what the girl will put in the box next to represent the month (one object per month). It reminds me of *The Important Book* by Margaret Wise Brown in that it says what is important to a child about a time of year.

A Flower Grows by Ken Robbins (Dial Books, 1990).

The book follows the growth of an amaryllis bulb through photos.

Planting a Rainbow by Lois Ehlert (Harcourt Brace Jovanovich, 1988).

This classic book shows bulbs in the ground before sprouting and when blooming.

Related *Early Years* Blog Posts

The following information is from "Planting This Fall for Springtime Blooms," published August 29, 2009:

I'm planning a fall gardening activity now, before school starts, and the first step is to mark my calendar to buy spring flowering bulbs to plant in fall. We sing a song before and after planting bulbs, and all winter long while wondering if the bulbs really will sprout. Act out the following song while you sing it, to the tune of the traditional song "Jack in the Box."

> *Spring flowering bulb, (children curl face down on floor, hiding face)*
> *So safe in the ground,*
> *Way down inside, your little dirt mound, (hands curve over the head)*
> *Spring flowering bulb so quiet and still,*
> *Won't you sprout up? (heads up and jump up, stretch arms up high)*
> *Of course I will!*

Read more at *http://nstacommunities.org/blog/2009/08/29/planting-this-fall-for-spring-time-blooms*.

Additional related blog posts:

"Planting Flower Bulbs in the Fall," published September 15, 2008: *http://nstacommunities. org/blog/2008/09/15/planting-flower-bulbs-in-the-fall*.

"Spring Flowering Bulbs Planted Where They Can Be Seen," published February 12, 2009: *http://nstacommunities.org/blog/2009/02/12/spring-flowering-bulbs-planted-where-they-can-be-seen*.

Chapter 35

Safe Smelling[36]

BEFORE YOU BEGIN THIS ACTIVITY:

Remember that each activity is not a stand-alone science or engineering curriculum. Activities are small steps in a journey of science inquiry, as discussed in Chapter 1. Your students will learn more about this concept and about the nature of science if you use this activity as part of an ongoing exploration of a question, a concept, or a topic being investigated by your class. Ask yourself, "What should come before this and what should come after?" Refer to Table 1.1 (pp. 15–39) to find other activities from *The Early Years* column that address the same concepts.

What smells do your children notice? I've heard students comment on the smell of blooming flowers, the lunch menu, and the gum that their teacher is not-so-surreptitiously chewing. Nothing smells better than a newborn baby's neck—that's my personal opinion. Children may prefer the smells of popcorn and cotton candy. Using the sense of smell, animals find food or a mate and detect the presence of predators; their survival depends on this. In a discussion on using our sense of smell to keep us safe, some children may relate experiences of smelling something burning. Identifying the five senses and corresponding sense organs builds toward understanding the disciplinary core idea LS1.D: Information Processing: "Animals have body parts that capture and convey different kinds of information needed for growth and survival. Animals respond to these inputs with behaviors that help them survive" (NGSS Lead States 2013).

Children know that smells come through the air from matter and that they detect smells with their noses, even though the mechanism—tiny particles of matter moving into the air and into our noses—can't be seen. Doing a sense-of-smell activity can be a way to introduce science time (laboratory) safety rules. In a science laboratory, adults and children must assume that no material is safe for contact through the mouth, eyes, nose, or skin. Although an early childhood classroom is not a real laboratory, noses are sensitive and can be irritated if children inhale food material such as vinegar or black pepper, so real safety procedures should be taught and

---

[36] This column entry was originally published in *Science and Children* in October 2009.

practiced (Roy 2007). *CAUTION: Check for allergies, and avoid using powders and soaps in the smelling activity.*

Young children can be expected to forget the rules, impulsively grab a container and take a great big whiff. To minimize this problem, teach the wafting technique of safe smelling: hold the container away from your face and wave your hand toward you, scooping the air and pushing it toward your nose. By waving part of the smell toward our nose and taking small, short sniffs, we protect our nose. It's easier for young children to follow this practice if another person holds the container while they waft and sniff.

Children may assume that the whole world experiences smell the way they do—a useful opportunity to introduce the concept of fact versus opinion. By looking at animals' noses and testing for favorite smells, young children learn that their sense organ for smell has the same function as other mammals, but preferences vary. Focus children's attention on the variety in noses by making a nose book in which photos of animal noses from magazines are covered by a second blank page with a hole cut in it, revealing just the nose. The noses are viewed first through the hole while children guess whose nose, and then they open the page to find out! Ask the children, *What smell do you think this animal might enjoy?* Children may answer according to their own preferences ("The bear likes to smell pizza!"), their observations (e.g., a dog sniffing the ground), or their stereotypical ideas (e.g., monkeys like to smell bananas). These statements can lead into a useful discussion about how we know and how we can find out. Learning about how animals use their external parts to survive and meet their needs is part of the *Next Generation Science Standards* grade 1 performance expectation 1-LS1-1 (NGSS Lead States 2013).

References

NGSS Lead States. 2013. *Next Generation Science Standards: For states, by states.* Washington, DC: National Academies Press. *www.nextgenscience.org/next-generation-science-standards.*

Roy, K. 2007. Scope on safety: When is a classroom a laboratory? *Science Scope* 30 (7): 74–75.

Activity: Do You Smell What I Smell?

Objective

To experience using our sense of smell and notice the variety in favorite smell choices

Materials

- Four opaque plastic or paper cups (Cover any pictures because they influence children's thinking.)

- Squares of thin cloth, such as dark nylon stockings or cheesecloth, or paper towels, to cover cups

- Rubber bands to hold the cloth over the cup

- Half a lemon

- Cinnamon sticks (To intensify the smell, store in ground cinnamon for several days or grate the stick with a grater or sandpaper to get a fresh surface.)

- Half an onion

- One-quarter cup of coffee beans

- Pictures of a lemon tree, a cinnamon tree, a coffee tree, and a growing onion

- Mirrors (one per child if available; handheld mirrors work well)

- Drawing materials

Teacher Preparation

Set up four opaque cups with safely smellable foods inside (lemon half, cinnamon sticks, onion half, coffee beans), covered with a thin cloth or paper towels held in place with a rubber band.

Procedure

1. Talk about the sense of smell while looking in a mirror and sniffing.

2. Have your students look in mirrors and describe the structure of their noses. To get them started, ask, *Do you have any holes in your nose?* Surprisingly, they often answer, "No."

3. Ask your students to draw their noses (it's easiest to begin with nostrils, which are a familiar oval and can be centered on the page). They may want to draw a complete face.

4. Ask your students, *How far away can you smell something?* Teach the wafting technique of smelling by waving part of the smell toward the nose and taking small, short sniffs. *CAUTION: Remind students to never smell directly from a container.*

5. In a group, play a smell guessing game using the four covered cups, beginning with the lighter-scented lemon, then cinnamon, then onion, and ending with the strongest smell, coffee beans. Have the children hold the container for the person next to them while they waft and sniff, to help them remember to sniff at a distance from the container.

6. Children can name their guesses as they smell, even though first sniffers may influence later ones. Even familiar food smells are surprisingly difficult to name when taken out of context. Reassure students that they may name the smell in any language. Depending on the home culture, children may or may not be familiar with the foods or the English names.

7. After every child smells a cup, remove the cloth to reveal the food inside. Then show the photo of the cultivated plant, showing how food comes from living plants.

8. Create a tally chart showing the food items in the smell guessing game. You can make one yourself, use the one in Figure 35.1, or download the chart in color from NSTA Connections at *www.nsta.org/elementaryschool/ connections/200910EarlyYearsFavoriteSmellsChart.jpg*. Review each picture with the children to be sure they connect the pictures with the foods they smelled. Have the children mark their favorite smell with a tally mark. Ask, *Does something that smells good to me always smell good to other people?* (Surprisingly, some children do like onion.) Then count up the number of tally marks for each food on a chart.

Children may name the dish or place they have smelled in association with the food, saying, "Oranges or lemonade," "cereal or cookies or sugar," "soup or chicken," and "Starbucks" rather than lemon, cinnamon, onion, and coffee. Smells create strong memories, linking us to past events or places.

FIGURE 35.1. Sample tally chart for the smell guessing game

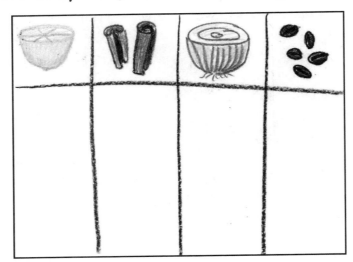

Teacher's Picks

Publications

What Can I Smell? by Sue Barraclough (Raintree, 2005).

Opening with the question, "What is your favorite breakfast smell?" this book invites discussion of familiar smells.

Two Eyes, a Nose and a Mouth by Roberta Grobel Intrater (Scholastic, 1995).

In a book full of photographs and rhyming text celebrating the variety in human faces, one page with repeated photos of just one face catches our attention, asks us to "imagine how dull the world would be, if everyone looked like you or me" and reminds us "the variety is just fine." Young children will enjoy pointing to the part of their bodies that they sense smells with or see, hear, or taste with.

Dog Breath: The Horrible Trouble With Hally Tosis by Dav Pilkey (Blue Sky Press, 1994).

Young children may not understand the title's play on words but they will get the humor of a dog with smelly breath saving the day. Ask your class, *When is our sense of smell useful?*

Smelling Things by Allan Fowler (Rookie Read-About Science; Children's Press, 1991).

An easy reader introduction to the sense of smell. Fowler's books pair simple, pertinent details about the topic with informative photographs.

The Happy Day by Ruth Krauss, illustrated by Marc Simont (HarperCollins, 1949).

Children can guess what the animals are smelling, but they will be surprised!

Website

The Herb Society of America
www.herbsociety.org

The "About Herbs" section of this website includes a beginner's guide and individual profiles of herbs.

Related *Early Years* Blog Posts

The following information is from "Favorite Smells—Stories and Activities," published October 1, 2009:

> *I love the way 2-year-olds inexpertly sniff, to sense an odor. They crinkle up their nose and snort, or gasp, and blink their eyes, not quite putting it all together to inhale through their nose. Yet they have an expert sense of smell—nothing comforts them like their favorite "lovey," a much worn toy or blanket that has achieved a certain smell. Scientists study the way smells affect people and our perceptions of smells. In my years of investigating smells with children, I have never had a student who was allergic to any of these foods: lemons, cinnamon, onions, and coffee beans. There is always a first time, so I check every class.*

Read more, including a recipe, at *http://nstacommunities.org/blog/2009/10/01/favorite-smells.*

Additional related blog post:

"Children Learn 'All About Me' While Using Science Tools," published September 11, 2010: *http://nstacommunities.org/blog/2010/09/11/children-learn-%E2%80%9Call-about-me%E2%80%9D-while-using-science-tools.*

Chapter 36

Helper Hats[37]

BEFORE YOU BEGIN THIS ACTIVITY:

Remember that each activity is not a stand-alone science or engineering curriculum. Activities are small steps in a journey of science inquiry, as discussed in Chapter 1. Your students will learn more about this concept and about the nature of science if you use this activity as part of an ongoing exploration of a question, a concept, or a topic being investigated by your class. Ask yourself, "What should come before this and what should come after?" Refer to Table 1.1 (pp. 15–39) to find other activities from *The Early Years* column that address the same concepts.

We often use the term *helpers* in early childhood education, as students learn about their community, beginning with home and family and expanding to the broader community—nation, continent, and world. Scientists use special equipment to do their work, including special clothing. Special clothing is worn by "community helpers" such as police officers, nurses, firefighters, cafeteria workers, dentists, and waste management workers as they do their jobs. The special clothing allows workers to perform their jobs and be safe. There may be jobs that are specific to or especially important for your community, such as volcanologists or lifeguards.

Children enjoy dressing up as community helpers because those are important jobs involving public safety. The workers have what seem to be "special powers" such as strength and knowledge, and they wear interesting equipment. For example, firefighters wear a lot of equipment to help them survive the sometimes extreme conditions of their job.

Exploring how hats help community workers do their jobs can be a way to introduce the idea of how the shape or form of an object is related to how it functions. In the natural world, the form of a plant or an animal's body can help us understand how it functions to ensure its survival and reproduction. If your hat is pointy, small for your head size, and brightly colored, what job are you dressed and adapted for? What if your hat is soft and covers your head closely, or is hard and has a short brim in the front and a longer brim in the back? It is hard to describe a hat fully, but children

---

[37] This column entry was originally published in *Science and Children* in February 2010.

easily point out features they see in photographs that make a hat a good tool for doing a particular job. Getting to try on hats is part of learning about these special tools.

Children can investigate how well community helpers' hats are suited for the job. What might your class investigate? Ask which clown hat gets the most smiles (the function of a clown hat is to elicit laughter, and they could count the number of smiles). To compare and contrast, you could ask why a clown hat is not appropriate for a firefighter. You could also ask which hat best identifies members of a team, which hat allows the wearer to carry heavy loads, which hat gives the best protection from the Sun or cold, or which hat gives the best protection from a spray of water. Asking questions and planning investigations are part of the science and engineering practices described in *A Framework for K–12 Science Education* (*Framework*; NRC 2012), a foundational document for the *Next*

Generation Science Standards (NGSS Lead States 2013). Investigating how people design and use hats introduces the relationship between "form" and "function." As stated in the *Framework*, "Exploration of the relationship between structure and function can begin in the early grades through investigations of accessible and visible systems in the natural and human-built world" (p. 97). Don't forget to ask your students, *What shape should a teacher's hat be?*

References

National Research Council (NRC). 2012. *A framework for K–12 science education: Practices, crosscutting concepts, and core ideas.* Washington, DC: National Academies Press. *www.nap.edu/catalog/13165/a-framework-for-k-12-science-education-practices-crosscutting-concepts*.

NGSS Lead States. 2013. *Next Generation Science Standards: For states, by states.* Washington, DC: National Academies Press. *www.nextgenscience.org/next-generation-science-standards*.

Activity: Will It Protect Me From the Water?

Objective
To explore the relation between a hat's structure and function

Materials

- Hats for three jobs (e.g., clown hat, sailor's knit watch cap, firefighter helmet)

- Pictures or photographs of people doing their jobs while wearing the hats (See the Morris book listed in the "Resource" section, or do an internet search for images using the hat name.)

- Writing materials (*Note:* Use dry-erase markers to write on page protectors with photo inside.)

- A mannequin head (make one by stuffing a bag with newspaper, tying it closed, and putting it on a stick stuck in the ground; or use a large doll head) or upside-down bucket

- Small cup of water for each child

Procedure

1. Show three hats and a picture of someone wearing each type of hat, one at a time, and ask the children to talk about each one. Pass the hats around so children can feel them and try them on. To begin a discussion, ask, *What shape is this hat?* and *Why do you think people wear a hat shaped like this?* and record the answers as documentation of the activity and thinking. Older children (writers) can record their own answers (see Table 36.1 for sample responses). It will be easier for young children to answer the questions if the three hats are quite different from each other and are specially designed for a particular job.

2. Have children talk about how each hat will function. *How does it stay on? How does it feel when you have it on? Will it keep you dry?*

3. Tell the children that when each of the three workers do their jobs, they may have water splashed on them—clowns spray each other with water from squirters, sailors get sprayed by the waves, and firefighters spray water from hoses on fires.

4. Ask, *Which hat will protect the worker the best from getting wet?* After a discussion, have the children try it by pouring water from a small container onto the top of each hat (supported on a mannequin head or bucket) to see where the water goes. Young children are able to do this in such a way that it exhibits what we want them to discover. They can be surprisingly careful about pouring and interested in observing and pointing out the flow. When every child pours there are multiple opportunities to observe, and

they can see that the flow is not chance but controlled by the hat structure. Because of their prior experience getting wet with rain, many children are not surprised that the clown hat gets wet, or when the water pours over the sailor's knit watch cap and on to the "head" but they are interested in the way the water flows over the curved part of the firefighters' hat and across the brim. After testing the protection the hats provide from water, the children are likely to say, "The firefighters' hat won!"

5. Ask the children to tell why the firefighter's hat provided the most protection from getting wet and they are likely to say that the brim kept the water off the head: "It rolled off."

Extension

Another function of the firefighter helmet that can be investigated is its ability to protect the head from falling objects. Children can wear a real or pretend helmet and drop a soft object, such as a foam block, on their own head. The sou'wester hats worn by fishermen have a similar shape to firefighter helmets but are made of different material. Have the children try on an actual sou'wester hat, or view a photograph, and then talk about how it is like and how it is different from a firefighter's helmet. Older children can make a Venn diagram to show how the hats' form and function are related.

TABLE 36.1. Sample student responses in helper hats activity

Type of hat	Response to "What shape is the hat?"	Response to "Why would a person wear a hat like this?"
Clown hat	• It's a triangle. • Pointy! • Kind of a small cone.	• It's got colors. • It's funny!
Sailor's knit watch cap	• It's round. • It's soft.	• To keep warm. • So my ears won't be cold.
Firefighter helmet	• Round! • It has a flat part (brim) and a strap.	• To keep the fire off. • To be safe. • To stay dry.

Resource

Morris, A. 1993. *Hats, hats, hats*. New York: HarperCollins.

Teacher's Picks

Publications

Zoe's Hats: A Book of Colors and Patterns by Sharon Lane Holm (Boyds Mills Press, 2009).

After seeing Zoe try on different hats (including some unexpected ones), readers can name their favorite at the end on a page showing all of the hats.

Firefighters A to Z by Chris L. Demarest (McElderry Books, 2003).

This book has been a favorite of one of my students for several years. The alphabet format includes firefighting vocabulary while informing readers about the work of firefighters.

The Wind Blew by Pat Hutchins (Simon & Schuster, 1974).

Ask your children, *Would the wind be able to blow your hat? Why or why not?*

Caps for Sale by Esphyr Slobodkina (William R. Scott, 1940).

This old favorite has been republished in many formats and is available in English and Spanish. Children can count the number of caps on the seller's head.

Website

Treasures of American History: American Hats
http://americanhistory.si.edu/treasures/american-hats

This website from the Smithsonian's National Museum of American History has photographs of a sampling of hats worn by Americans throughout the centuries.

Related *Early Years* Blog Posts

The following information is from "Early Education in Engineering and Design," published October 12, 2013:

> *I love to try to understand the thinking process children are going through as they work to solve a problem involving objects in space—putting a puzzle together, making a system to carry toys, figuring out which arm to put into a coat sleeve and balancing blocks. What is the motivation and*

what is the goal for a child working in the housekeeping area, trying to figure out how to store all of the play food in the play microwave when pieces keep falling out? As children look for solutions to these kinds of problems, we can encourage them to think about what the difficulty is and explore possible solutions, perhaps more than one solution. Designing a best solution for a problem is a process that continues, with solutions being tested, reconsidered, and redesigned.

Read more at *http://nstacommunities.org/blog/2013/10/12/early-education-in-engineering-and-design*.

Additional related blog posts:

"Exploring Form and Function With Hats: Books About Firefighters," published February 4, 2010: *http://nstacommunities.org/blog/2010/02/04/exploring-form-and-function-with-hats-books-about-firefighters*.

"Found Materials + Engineering Process = Toy," published April 26, 2012 *http://nstacommunities.org/blog/2012/04/26/found-materials-engineering-process-toy*.

Chapter 37

Building With Sand[38]

BEFORE YOU BEGIN THIS ACTIVITY:

Remember that each activity is not a stand-alone science or engineering curriculum. Activities are small steps in a journey of science inquiry, as discussed in Chapter 1. Your students will learn more about this concept and about the nature of science if you use this activity as part of an ongoing exploration of a question, a concept, or a topic being investigated by your class. Ask yourself, "What should come before this and what should come after?" Refer to Table 1.1 (pp. 15–39) to find other activities from The Early Years *column that address the same concepts.*

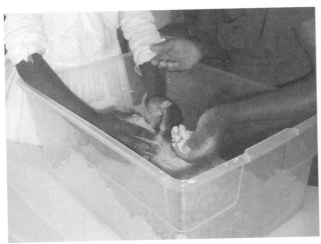

Children playing in damp sand invariably try to make a tower or a tunnel. At a sensory table filled with dry sand, they fill and pour sand from containers—and they do the same when the table holds water. By providing experiences with a variety of materials, alone and together, teachers set up the conditions for children to learn through their senses and ensure that a class approaches a topic with a common set of experiences to build on. By using measurement and comparison in science activities, children discover that scientists use math to learn about the world.

To prepare children to discuss building structures using sand, explore the properties of dry sand and water separately, in separate tubs or sequentially in a sensory table—water one week (or month) and sand the next. Children need time to get more than one turn at the table to interact with the material with a variety of tools, to talk about their work with other students and the teachers, and to make plans to try an idea. Vary the tools—cups, spoons, bottles, berry baskets, sieves, marbles, dolls, funnels, tubes, plates—and children will vary the

---

[38] This column entry was originally published in *Science and Children* in March 2010.

National Science Teachers Association

task by filling bottles, washing dolls, covering up cups, burying marbles, and so on. The tubs become a center for exploration of the nature of the material. Post a piece of paper or clipboard nearby for recording dictated thoughts or writing and drawing as a way for children to record their thinking to use later in a short discussion in circle time. Ask open-ended questions such as, *Can you pour the water/sand? Tell me how the water/sand feels on your hand. How are water and sand the same/different? Where have you seen sand in nature? What is water/sand made of?* Although they probably cannot answer this last question, it introduces the idea that all materials are made of something. Learning about the properties of materials builds toward disciplinary core idea 2-PS1: Matter and Its Interactions (NGSS Lead States 2013), and Constructing Explanations (for science) and Designing Solutions (for engineering) are part of the science and engineering practices (NRC 2012).

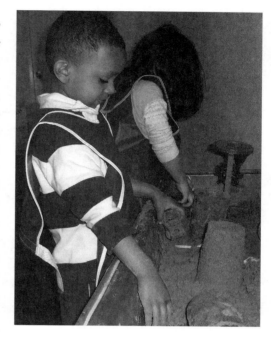

After exploring the water and sand separately, children will be eager to mix them together. Another period of exploration will teach children that wet sand does not behave like dry sand or like water. Continue recording students' observations and ideas to refer to in discussion later. When the experience has been shared by all in the classroom, hold a class discussion about building with sand, and why sand will not pile up unless it is wet, to encourage children to continue thinking about and sharing their ideas to solve problems or provide answers—something scientists and engineers do. Even young children are able to "partake in science discussions that would include participating, negotiating, taking turns, and listening to others" (Sander and Nelson 2009). Children reveal their prior experience building sand castles when they say that they need wet sand to build a structure.

Of course, children will want to test the ideas! Continue asking them questions. Children will use what they have learned through playing with the dry sand and water to solve the problem of how to build a tower or tunnel with sand and water— the work of engineers.

References

National Research Council (NRC). 2012. *A framework for K–12 science education: Practices, crosscutting concepts, and core ideas.* Washington, DC: National Academies Press. *www.nap.edu/catalog/13165/a-framework-for-k-12-science-education-practices-crosscutting-concepts.*

NGSS Lead States. 2013. *Next Generation Science Standards: For states, by states.* Washington, DC: National Academies Press. *www.nextgenscience.org/next-generation-science-standards.*

Sander, J., and S. Nelson. 2009. Science conversations for young learners. *Science and Children* 46 (6): 43–45.

Activity: How High Can You Build With Sand?

Objective

To explore water, sand, the force of water-surface tension that holds sand grains together when wet, and wet sand as a building material

Materials

- Sensory table or tub(s)

- Water

- Sand, 25 lb or more

- Measuring cup

- Measuring tool for height—unit cubes or centimeter ruler

- Tools and toys for playing in sand or water

- Magnifying lens

- Writing materials for recording ideas, observations, and drawings

Procedure

1. Ensure sensory table access for every child for up to a week with water and with dry sand—and later, wet sand—before beginning the structured inquiry. Vary the tools and toys in the table. *CAUTION: Have children wash hands before and after playing in the sensory table.*

2. Record ideas about the material you overhear in the conversation around the sensory table and have the children draw or write about their experience with the material. Allow the sand to dry completely.

3. After all students have participated in the initial exploration period, bring the class together for a group discussion. Tell them that you are going to record their ideas as the class discusses a science and engineering problem—a problem for which they will use what they have learned about the nature of materials to design and build a tower or tunnel from sand.

4. Ask the students whether they think that they can build a small tower or tunnel using sand. After hearing their "Yesssss!" show them a tub of dry sand and have them try, one by one, to build and measure the height of a tower. Because they have had prior experience with the dry sand, they will not need much time to test its usefulness as a building material (but you may still have to set a short time limit).

5. Tell the class that you want them to talk about the problem and discuss ideas—that scientists and engineers talk out their ideas before trying them. Begin the discussion by reading the previously recorded observations and ideas and ask, *What can we do to make this sand work as a building material for our small structures?* List the ideas and choose "add water" as the simplest (and least messy) idea to try.

6. Have the children look at a pinch of dry sand in their hand and how sand grains behave, using a magnifying lens to look closely. Next, have them observe what happens when a drop of water is added. Comments such as "It sticks together" will support their plan to add water.

7. Depending on the age of the students, water can be added to dry sand "to touch" or in a more controlled way, by measuring and adding different amounts of water to containers with the same amount of dry sand. Begin with the ratio of 1:1 (this will be way too much water, but students initially think more will be needed) and work down from there. Have the students build towers, tunnels, and more structures using the now-wet sand. Have them record or dictate what they notice and, in the case of the measured wetness, compare the resulting tower heights. Support the students' investigation by asking questions such as, *Can a higher tower be built with wet sand?* and *How much water to how much dry sand makes the tallest tower?*

Children may want to try some of the more elaborate ideas for making dry sand into a workable building material. They may suggest using the dry sand to fill a hollow form or container, or adding liquid glue to the sand. Let them try these ideas out!

Teacher's Picks

Jump Into Science: Sand by Ellen Prager, illustrated by Nancy Woodman (National Geographic Children's Books, 2006).

This book's illustrations include close-ups of a magnified view of sand.

From Sand to Glass by Shannon Zemlicka, photos by Randall Hyman (Start to Finish; Lerner, 2004) and *Sand to Glass* by Inez Snyder (Welcome Books; Children's Press, 2005).

With a table of contents, a glossary, and an index, *From Sand to Glass* teaches early readers how to use nonfiction while telling the interesting story of how sand becomes glass, another familiar but very different material. The second book is similarly structured but with less text.

Sand by Pam Miller, illustrated by Rick Stromoski (Rookie Readers; Children's Press, 2000).

With 61 words on the word list, this early reader covers the places we might see sand and the conditions we find it in—beach, desert, in piles, wet, and dry.

Sand Castle by Brenda Yee, illustrated by Thea Kliros (Greenwillow Books, 1999).

On a sandy beach a girl accepts the help of one child after another to build a sand castle, working with *wet* sand.

Super Sand Castle Saturday by Stuart J. Murphy, illustrated by Julia Gorton (Mathstart; HarperCollins, 1999).

Larry the Lifeguard starts a sand castle–building contest, and the children measure with shovel lengths, steps, and inches to determine the tallest, longest, and deepest.

Related *Early Years* Blog Posts

The following information is from "Soil Erosion in Miniature," published May 13, 2013:

> *There are not many fiction or nonfiction books for young children that include a discussion about soil erosion, or erosion in general. If your class becomes interested in learning about erosion, they can write their own story, with your help, and illustrate it with photos or drawings of*

places where they saw soil washed away by water or blown away by wind. Children can make a model of a neighborhood in sand in a plastic tub or in the sandbox and then make it "rain" to demonstrate erosion. They can also create a book about erosion. Getting the Most Out of Picture Books, *a video series on "dialogic reading" from the Getting Ready to Read pages of the National Center for Learning Disabilities, shows how to get the most out of shared reading.*

Read more at *http://nstacommunities.org/blog/2013/05/13/soil-erosion-in-miniature.*

Additional related blog posts:

"What Science Happens in Your Sandbox?" published April 19, 2013: *http://nstacommunities. org/blog/2013/04/19/what-science-happens-in-your-sandbox.*

"Preschool STEM," published March 1, 2010: *http://nstacommunities.org/blog/2010/03/01/ preschool-stem.*

Chapter 38

Inquiry at Play[39]

BEFORE YOU BEGIN THIS ACTIVITY:

Remember that each activity is not a stand-alone science or engineering curriculum. Activities are small steps in a journey of science inquiry, as discussed in Chapter 1. Your students will learn more about this concept and about the nature of science if you use this activity as part of an ongoing exploration of a question, a concept, or a topic being investigated by your class. Ask yourself, "What should come before this and what should come after?" Refer to Table 1.1 (pp. 15–39) to find other activities from The Early Years column that address the same concepts.

Play and science inquiry are essential parts of early childhood programs. Imaginative play, unscripted yet guided by children's own rules, allows students to use their imagination and develop self-regulation, symbolic thinking, memory, language, and social skills, as well as construct their knowledge and understanding of the world. Play can reflect what children learn while engaged in science inquiry. Like play, science inquiry helps children make sense of their world and appreciate the work of scientists.

Making observations, asking questions, planning investigations, gathering and sharing data, interpreting data, considering alternate explanations, discussing patterns and relationships, solving problems, and asking new questions are all part of science inquiry. The idea of there being one or "the" scientific method should be discarded, because science inquiry is not a one-way process but continues as investigations lead to understanding, raising new questions, and creating more investigation. Learning about and participating in science inquiry involves children in the practices

---

[39] This column entry was originally published in *Science and Children* in September 2010.

of science and engineering (NRC 2012) and deepens their understanding of the nature of science (NGSS Lead States 2013).

Young children may be susceptible to stereotypical images of scientists and what scientists do. One way to teach children about the breadth of science in their lives and to think inclusively about scientists is to cut out and laminate pictures of scientists at work in many settings (Ashbrook 2005). The photos should show scientists looking, drawing, writing, measuring, working with collections, talking with each other, using special equipment to work in various environments, and describing their work.

Being able to tie the word *scientist* to a particular person may also help children understand the work of scientists. Invite a scientist into the classroom for a "play-date." If a scientist is not available, use a few simple props to become one specific scientist yourself, avoiding the stereotypical "scientist in a lab" image so children will learn that scientists work in many settings. Young children often use imaginative play to make meaning of what they are learning. Dress up as figures such as Leonardo da Vinci (artist, scientist, and inventor), Jane Goodall (primatologist and anthropologist), or the Chinese empress Si Ling-chi, who, according to legend, developed the process for making silk fabric.

As you impersonate a scientist, maintain a friendly manner while giving direct instruction on the science investigation. As a "visiting scientist," you will teach the children new vocabulary, factual science content, and process skills to incorporate into their inquiry. Using props related to the area of study will make you the scientist even if you are a different gender or body type. Later, you can present students with new props representing other scientists and science work and have them think about what kind of scientist would use those kinds of tools and for what purpose.

References

Ashbrook, P. 2005. What can young children do as scientists? *Science and Children* 42 (9): 24–26.

National Research Council (NRC). 2012. *A framework for K–12 science education: Practices, crosscutting concepts, and core ideas.* Washington, DC: National Academies Press. *www. nap.edu/catalog/13165/a-framework-for-k-12-science-education-practices-crosscutting-concepts.*

NGSS Lead States. 2013. Appendix H—Understanding the scientific enterprise: The nature of science in the *Next Generation Science Standards*. In *Next Generation Science Standards: For states, by states*. Washington, DC: National Academies Press. *www.nextgenscience.org/ next-generation-science-standards*.

Activity: A Scientist Visits

Objectives

- To introduce a scientist (through a classroom visit from a real scientist or by the teacher dressing as a historical or living scientist)

- To learn science content about a topic

- To relate classroom science inquiry to the work of actual scientists

Materials

- Photographs and biographies of scientists (Look in magazines such as *National Geographic* for images of scientists in the field and in the lab, both men and women and of various ethnicities.)

- A local scientist (if available) or costumes (adult and child-size) and props for the scientist you choose to represent

Procedure

1. Using photographs of scientists, talk with students about how the science inquiry they do in the classroom is similar to working scientists' activities (e.g., drawing or representing what they see, asking questions, and trying to find answers).

2. Discuss the equipment needed by scientists working in various fields.

3. Invite a local scientist to visit your classroom to teach about a topic or concept your class is learning about through inquiry. This will take some legwork. Do any of the children in your school have parents who are scientists and can talk to the class about their work? Check into availability for classroom visits with an environmental technician or scientist from your local water treatment facility, an arborist from a tree maintenance company or state office, a soil scientist from the Soil Science Society of America (through the Ask a Soil Scientist program; see "Resources" section), a naturalist from a local nature center, or a physicist from a nearby college.

4. Have the children brainstorm questions they want to ask the visiting scientist.

5. If no actual scientist is available, become a historical or modern scientist by dressing up with the appropriate costume. Research facts about the scientist and his or her work that you can convey in a "visit" of about 15 minutes, including answers to the questions the children plan to ask. On the day of the visit, put on the costume and props out of sight and return in character. The children will know that you are still the teacher, but they may interact with you as though you are the scientist you represent.

6. Have the visitor (or you, the costumed scientist) talk with the students about his or her work and answer questions about the work. Ask questions of the students about their science inquiry work.

7. After the visit, have the class list information they learned or what they noticed about the scientist, and have them write thank-you notes to the scientist that mention something they learned.

Extension

Continue teaching about scientists by reading aloud age-appropriate biographies of scientists. Add child-size costumes and props to an "imaginative play" area so the children can reenact the visit and reveal their understanding of the scientist's work through their play. Support complex storylines by modeling how to ask a question, suggesting discussion with other scientists, and asking what questions the children or scientists have and what plans they have for finding out.

If you are concerned that people observing your class will only see the "play" and not the resulting intellectual development, be sure to document the learning by having children draw, write, and tell about the meaning they make as they play and how it relates to the class's science inquiry and the direct instruction that goes along with inquiry. Share with parents and administrators a copy of the position statement of the National Association for the Education of Young Children on developmentally appropriate practice in early childhood programs, which emphasizes the importance of play in early childhood (NAEYC 2009).

Reference

National Association for the Education of Young Children (NAEYC). 2009. *Position statement: Developmentally appropriate practice in early childhood programs serving children from birth through age 8.* Washington, DC: NAEYC. *www.naeyc.org/files/naeyc/file/positions/PSDAP.pdf.*

Resources

Smithsonian Science News. Meet our scientists. *http://smithsonianscience.si.edu/category/meet-our-scientists.*

Soil Science Society of America. K–12 Soil Science Teacher Resources. *www.soils4teachers.org/ask.*

Teacher's Picks

Publication

Scientists at Work by Susan Ring (Capstone Press, 2006).

This book helps young readers understand the many types of jobs that are available in the field of science. It includes photographs of scientists at work and descriptions of their fields of specialization.

Website

SCI4KIDS Cool Careers
www.ars.usda.gov/is/kids/CoolCareers/coolestcareers.htm

This website from the U.S. Department of Agriculture, Agricultural Research Service, has photos and descriptions of different types of scientists (e.g., biologist, botanist, chemist, engineer, nutritionist).

Related *Early Years* Blog Posts

The following information is from "Celebrate Pink!" published June 23, 2009:

> *What can we do to encourage all children to think of themselves as capable and support an interest in science? Remember to alternate saying "boys and girls" with "girls and boys"? Monitor whom we call on, and call on girls 50% of the time? Be sure to display pictures of all kinds of people, including women, doing science? Many women scientists say they pursued a career in science because a teacher or other mentor encouraged them and believed in their capabilities. This idea is echoed in many of the life stories of women scientists on the "Women@NASA" web page (http://women.nasa.gov) of the National Aeronautics and Space Administration. Print out and post some of the photos to inspire and inform children and testify to women's presence in science.*

Read more at *http://nstacommunities.org/blog/2009/06/23/celebrate-pink.*

Additional related blog posts:

"What Is a Scientist? Resources for Young Children," published September 6, 2010: *http://nstacommunities.org/blog/2010/09/06/what-is-a-scientist-resources-for-young-children.*

"Preschool Play as Assessment Tool," published June 1, 2009: *http://nstacommunities.org/blog/2009/06/01/preschool-play-as-assessment-tool.*

"Raise Your Hand If You're a Scientist!" published June 19, 2009: *http://nstacommunities.org/blog/2009/06/19/raise-your-hand-if-youre-a-scientist.*

Chapter 39

Developing Observation Skills[40]

BEFORE YOU BEGIN THIS ACTIVITY:

Remember that each activity is not a stand-alone science or engineering curriculum. Activities are small steps in a journey of science inquiry, as discussed in Chapter 1. Your students will learn more about this concept and about the nature of science if you use this activity as part of an ongoing exploration of a question, a concept, or a topic being investigated by your class. Ask yourself, "What should come before this and what should come after?" Refer to Table 1.1 (pp. 15–39) to find other activities from The Early Years column that address the same concepts.

A new teacher once asked me, "What are the science process skills and practices, and how will I know if my students are learning how to use them?" We typically know children are learning when they are able to make sense of an object's materials or a situation that was previously a bit mysterious and communicate what they have figured out. But what about observing? In grades K–2, the focus is on "observations and explanations related to direct experiences" (NRC 2012, p. 34). The *Next Generation Science Standards* describe observation as part of the science practice Asking Questions and Defining Problems (NGSS Lead States 2013). Observation is something students have been practicing all their lives. As with all skills, children become more capable as they use these skills with adult guidance. Teachers can ask, *What is your nose telling you? What are your eyes telling you?* and *What does each of your senses tell you?* Challenge children to tell you what they see, instead of what they think they should see, by doing a color challenge with illustrations of a green fire truck, a pink snowflake, and a blue Sun.

For an activity about using and understanding our sense of smell (see Chapter 35, "Safe Smelling," p. 208), I put pungent food—lemons, cinnamon sticks, onions, and coffee beans—in small opaque containers and covered the tops with paper towels held on by rubber bands. What I didn't do was cover the sides of the containers. They were decorated with pictures of the fruit flavors of the yogurt that

---

[40] This column entry was originally published in *Science and Children* in October 2010.

came in the containers. As the children wafted the lemon scent toward their noses, they guessed blueberry because that was what they thought was supposed to be in the container, not what their noses smelled.

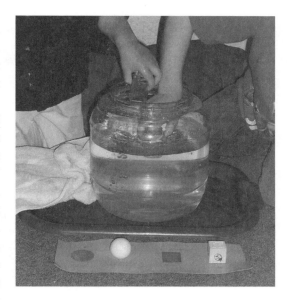

After covering up the pictures on the sides of the yogurt containers, I asked the children, "What does that smell like to you?" rather than "What is in the container?" Some answers were "gum" and "chicken when it's cooking" (for cinnamon and onion). The kinds of questions teachers ask affect students' level of thinking (Blosser 2000). Asking questions that can be answered with a simple statement denies children the chance to make detailed observations. Ask questions that are open-ended, giving the students time to describe what they observe. The length of time you wait after asking a question is also important (Rowe 1986). Children need time to formulate their answers, to find the language, and sometimes to find the self-assurance to speak up.

References

Blosser, P. E. 2000. *How to … ask the right questions.* Arlington, VA: NSTA Press.

National Research Council (NRC). 2012. *A framework for K–12 science education: Practices, crosscutting concepts, and core ideas.* Washington, DC: National Academies Press. *www. nap.edu/catalog/13165/a-framework-for-k-12-science-education-practices-crosscutting-concepts.*

NGSS Lead States. 2013. *Next Generation Science Standards: For states, by states.* Washington, DC: National Academies Press. *www.nextgenscience.org/next-generation-science-standards.*

Rowe, M. B. 1986. Wait time: Slowing down may be a way of speeding up! *Journal of Teacher Education* 37 (1): 43–50.

Activity: Making and Observing Bubbles

Objective
To make bubbles and make observations about their shape

Materials

- Feather

- One straw and small cup (3–9 oz) for each child

- Water

- Large, tall (20–30 cm), clear plastic container (e.g., empty pretzel or cheese ball container)

- A small clear plastic container with a circular opening (e.g., empty spice jar)

- A small clear plastic container with a square opening (e.g., craft store bead container)

- A Ping-Pong–size ball and a cube block (about the same size; alphabet blocks work well)

- Towel or tray to catch spilled water

- Drawing and writing materials

Procedure

1. Use shape names to talk about two-dimensional objects (e.g., "the plate shaped like a circle" or a "rectangle-shaped tabletop") and three-dimensional objects (e.g., "the globe that is a sphere—shaped like a ball" or "the cube-shaped block") when referring to classroom objects.

2. Illustrate the concept of a *gas* by demonstrating deep breaths, puffing your cheeks with air, and moving a feather by blowing on it.

3. Give each child a straw to blow through, first onto their hands, and then into a small amount of water in their cups to feel that air is "something"— it is matter. Ask, *What did the air look like when it went through the water? What did you see?* It is difficult to see the shape of an individual bubble in a stream of bubbles, so don't be surprised to hear many descriptions.

4. Have the children examine the small container with the circular opening.

5. Fill the large container about three-quarters of the way with water.

6. Demonstrate how bubbles can be made by putting the small container entirely under the water, with the open end down, and then tipping it up to release the air one bubble at a time. Have the children describe and draw what shape bubble they observed, with the ball and square block in view for children to point to if they don't remember the terms. Once is never enough—give each child a turn making bubbles.

7. Have the children examine the small container with the square opening, ask for their predictions of what shape bubbles will be made, and repeat

the process of making bubbles and having the children say and record their observations. Most young children will predict that the bubbles will have a square (cube) shape and will say that they observe cube-shaped bubbles.

8. As the children take turns making the bubbles themselves and observing while others make the bubbles, a few children will correct themselves and say, "The bubbles are like a ball" (spherical).

9. Discuss the children's observations using their drawings. Tell them, *It is okay to have different ideas. We can continue making bubbles and also use other ways to make bubbles to see if we can agree on what shape bubble comes from a square opening.* Tell them that in science, observations may need to be made over and over again, for a long time.

It may take many sessions of bubble making, in water or with bubble solution using square-shaped bubble wands made from pipe cleaners, before all children will accept what their visual sense tells them (sphere-shaped bubbles) rather than what they believe they should be seeing (cube-shaped bubbles).

Teacher's Picks

Publications

Pop! A Book About Bubbles by Kimberly Brubaker Bradley, with photographs by Margaret Miller (Let's-Read-and-Find-Out Science, Stage 1; HarperCollins, 2001).

This book teaches children all about bubbles, including the science behind them, and suggests some questions about bubbles to investigate

Cubes, Cones, Cylinders, and Spheres by Tana Hoban (Greenwillow Books, 2000).

The photographs in this book show children that the shapes listed in the title are found all around us. Using the photographs in this wordless book, have children name the shapes they see and teach them the names of unfamiliar shapes.

The Nature and Science of Bubbles by Jane Burton and Kim Taylor (Exploring the Science of Nature; Gareth Stevens, 1998).

This book explains how, why, and where bubbles are formed and describes their different uses and various appearances.

Website

Bubble Festival: Presenting Bubble Activities in a Learning Station Format
www.lhsgems.org/GEM132.html

This GEMS (Great Explorations in Math and Science) unit, written by Jacqueline Barber and Carolyn Willard, is from the University of California, Berkeley, Lawrence Hall of Science. Designed for grades K–6, it uses 12 tabletop bubbles-related activities to present exploratory lessons in math and science.

Related *Early Years* Blog Posts

The following information is from "Observing Closely—Bubbles!" published October 10, 2010:

> *Children with experience blowing bubbles may not be able to recall the shape of, or say the name of, the shape of all free-floating bubbles—a sphere. Bubble blowing is a good time to talk about the difference between two-dimensional "round" objects and three-dimensional "round" objects and to have children practice careful observation. Use familiar classroom objects such as balls and marbles, cube blocks, and boxes to compare with paper cutouts of the two-dimensional shapes (a circle and a square).*

Read more at *http://nstacommunities.org/blog/2010/10/10/observing-closely%E2%80%94bubbles*.

Additional related blog post:

"Preventing Misconceptions," published September 24, 2008: h*ttp://nstacommunities.org/blog/2008/09/24/preventing-misconceptions*.

Chapter 40

Investigable Questions[41]

BEFORE YOU BEGIN THIS ACTIVITY:

Remember that each activity is not a stand-alone science or engineering curriculum. Activities are small steps in a journey of science inquiry, as discussed in Chapter 1. Your students will learn more about this concept and about the nature of science if you use this activity as part of an ongoing exploration of a question, a concept, or a topic being investigated by your class. Ask yourself, "What should come before this and what should come after?" Refer to Table 1.1 (pp. 15–39) to find other activities from The Early Years column that address the same concepts.

Does curiosity provoke a question? Children (like adults) are curious about what is new. Galinsky (2010) discusses Dr. Laura Schulz's research on the role of young children's curiosity in learning from direct evidence. Dr. Schulz found that children are also curious when something contradicts what they expect (their prior understanding) about the world and when they see evidence that fails to distinguish among competing beliefs (when evidence doesn't make it clear why something works). Here's an example from my experience: Young children try repeatedly to blow bubbles with a bubble solution that has become too sudsy to sustain a large bubble. They know the solution "worked" previously and try blowing harder, dipping faster, and using a new wand, but they are not able to form bubbles or understand why.

Children often ask "about" or "what is" questions, such as, "What kind of seed is this?" Pearce (1999) says that young children often ask "Can I ..." questions when first stating investigable questions, such as "Can I blow a big bubble

---

[41] This column entry was originally published in *Science and Children* in December 2010.

with this tool?" He suggests that teachers can deepen those questions by asking children to rephrase into a quantifying question, such as, "How big a bubble can I make with this tool?" which also leads to learning about measuring. For students who are readers and writers, Pearce (1999; see also Workshop 4 of *Learning Science Through Inquiry* [Annenberg Learner 2015]) offers a worksheet with testable questions that contains sections for recording observations, recording questions, completing stems (e.g., "How can we...?"), and comparing one situation with another ("When comparing sunflowers with suet, which will attract the most birds?").

Teachers may not need to teach children to ask questions; instead, teachers should develop a safe place where questions can be voiced, observe children to see the questions in their actions, and develop a culture that appreciates and records questions. An investigable question is rare in the preschool years, but with questions so readily voiced, this is the time to begin making children aware of what they can and cannot answer through investigation. With a classroom culture of welcoming questions and providing time and resources to find out, the young child's "Why?" turns into a more complex question, such as, "Do roly-polies (also known as pill bugs) change as they grow the way butterflies do?"

Asking Questions and Planning and Carrying Out Investigations are two of the science practices first described in *A Framework for K–12 Science Education* (NRC 2012) and later incorporated into the *Next Generation Science Standards* (NGSS Lead States 2013). The National Science Teachers Association's 2014 position statement on early childhood science education affirms "that young children have the capacity for conceptual learning and the ability to use the skills of reasoning and inquiry as they investigate how the world works."

Although winter may seem an unlikely time to observe seed sprouting, it is actually a good time to do this because in many parts of the country classes stay indoors for longer periods and therefore will have time to water and observe. Developing understanding before the spring will prepare children to garden when the temperature allows. A class discussion on the needs of seeds to be able to grow will raise the need for water. Most children know this through hearing it or direct experience. But they may have different opinions on how much water is the right amount for best seed sprouting. Of course, "how much water" depends on many things: type of seed, humidity of the air, surface area of the container, whether the container has a cover, and if it is in sunshine or shade. Control these variables by using the same container, in the same place, with the same kind of seed, and vary just one thing—the amount of water.

References

Annenberg Learner. 2015. Learning science through inquiry: Support materials. [See the appendix for the "Question Search" handout in Workshop 4.] *www.learner.org/workshops/inquiry/support/index.html.*

Galinsky, E. 2010. Critical thinking. In *Mind in the making: The seven essential life skills every child needs*, 224–228. New York: HarperCollins.

National Research Council (NRC). 2012. *A framework for K–12 science education: Practices, crosscutting concepts, and core ideas.* Washington, DC: National Academies Press. *www.nap.edu/catalog/13165/a-framework-for-k-12-science-education-practices-crosscutting-concepts.*

National Science Teachers Association (NSTA). 2014. *NSTA position statement: Early childhood science education. www.nsta.org/about/positions/earlychildhood.aspx.*

NGSS Lead States. 2013. *Next Generation Science Standards: For states, by states.* Washington, DC: National Academies Press. *www.nextgenscience.org/next-generation-science-standards.*

Pearce, C. R. 1999. *Nurturing inquiry: Real science for the elementary classroom.* Portsmouth, NH: Heinemann.

Activity: How Much Is Enough?

Objectives

- To introduce the idea of investigating a question

- To investigate how much water is best for mung bean sprout growth through a fair test

Materials

- Writing and drawing materials

- Mung bean seeds (or another quickly sprouting and growing seed)

- Three clear 5–9 oz cups

- Label for each cup with a drawing depicting these words: "no water" (for an empty cup); "a little water" (for an identical cup with 1–2 cm water on the bottom); and "a lot of water" (for an identical cup filled two-thirds with water) (*Note:* Make the labels ahead or incorporate this as part of the activity.)

- Water

- Permanent marker

- Droppers (pipettes) (optional)

Procedure

1. Hold a class discussion on what makes a seed sprout. Entertain and record all ideas and then focus on the question of how much water seeds need to sprout "best." Children most likely define *best* by the presence of vigorous green growth. Tell the class that they can investigate the question by making a fair test of how much water is best. Children may want to record each step of this procedure using writing and drawing materials as the seeds sprout.

2. Show the labeled three cups and discuss the meaning of the words and pictures, to be sure the students all understand how much water goes in each cup.

3. Talk about how everything in a fair test is the same except for one thing. You are using the same size cup for each amount of water, the same kind of seeds, and the same amount of seeds. The cups will go in the same place (windowsill or desk), so they will be getting the same amount of light and be at the same temperature. Only the amount of water will be different, to test for the best amount of water for seed growth.

4. Have the children pour the appropriate amount of water into each cup: none, a little (1–2 cm), and a lot (two-thirds full).

5. Ask children in which cup they think the seeds will sprout and grow best, reminding them that scientists do not all have to agree. As a rough assessment of their prior knowledge about seed sprouting, have students drop a seed into the cup that they think is best for seed sprouting and growth. Children may select for other reasons—for example, to add to a cup that is empty or to follow a friend's lead.

6. After each student has a turn, have them count the number of seeds in each cup. Add seeds where needed, reminding students that for a fair test, each cup must hold the same number of seeds. Use a permanent marker to mark the water level so more can be added as some evaporates.

7. Have students check on the seeds every day for a week or two for signs of growth. Using droppers to make it easier to add small amounts, children

can add water up to the marked level if necessary. Record seed growth in drawings labeled with the date.

Spoiler alert! The seeds in the "no water" cup will not grow. The seeds in the "a little water" cup will sprout and send up a stem and set of second leaves. The seeds in the "a lot of water" cup will sprout, begin to grow, and then stop growing and ferment. Stop the fair test when children have determined a clear best amount of water for growth.

Discuss the results of the experiment while reviewing the drawings documenting growth. Wonder aloud about the results and encourage children to share their ideas. They may not realize that plants need air, just like people. Remind them that they cannot breathe underwater and tell them that many plants cannot either.

Teacher's Picks

Publication

"Inquiry on Board!" by Helen Buttemer, in *Science and Children* 44 (2): 34–39, 2006.

This article describes how Buttemer helps students identify variables, an understanding that is central to conducting a "fair test" in science.

Websites

Doing a Fair Test: Variables for Beginners
www.sciencebuddies.org/science-fair-projects/project_experiment_fair_test. shtml

This page on the Science Buddies website provides a brief description of what a fair test is in science experiments.

Understanding Science 101. Fair Tests: A Do-It-Yourself Guide
http://undsci.berkeley.edu/article/fair_tests_01

This page on the Understanding Science website provides a good explanation of the components of a fair test in science experiments.

Related *Early Years* Blog Posts

The following information is from "Seed Sprouting, Activity and Observation," published March 7, 2009:

> *It's fun for children to plant seeds in a special container, but it can be hard to remember to water them, leading to disappointment if the plants don't survive. Planting grass seed in some bare spots on any lawn is just as satisfying, perhaps more so because with time it will be hard to say which grass plant is the "one" they planted, and therefore they can claim the success of all.*

Read more at *http://nstacommunities.org/blog/2009/03/07/seed-sprouting-activity-and-observation*.

Additional related blog posts:

"Gardening Begins, Inch by Inch," published March 15, 2010: *http://nstacommunities.org/blog/2010/03/15/gardening-begins-inch-by-inch*.

"Science Activities in Early Childhood Prepare for a Lifetime of Learning," published March 30, 2009: *http://nstacommunities.org/blog/2009/03/30/science-activities-in-early-childhood*.

Chapter 41

Ongoing Inquiry[42]

BEFORE YOU BEGIN THIS ACTIVITY:

Remember that each activity is not a stand-alone science or engineering curriculum. Activities are small steps in a journey of science inquiry, as discussed in Chapter 1. Your students will learn more about this concept and about the nature of science if you use this activity as part of an ongoing exploration of a question, a concept, or a topic being investigated by your class. Ask yourself, "What should come before this and what should come after?" Refer to Table 1.1 (pp. 15–39) to find other activities from The Early Years column that address the same concepts.

Many early childhood programs are scheduled for just three hours in the morning, and in those that are full-day programs the time set aside for breakfast, snacks, and naps reduces the amount of available teaching time. Because of these time constraints and because young children need help with small-motor activities and self-help skills, teachers often prefer science inquiry experiences that are easy to set up, clean up, and store. Teachers also value science activities that include mathematics and literacy components—criteria easily satisfied in early childhood by including counting, measurement, looking for patterns, and having children document their work with drawings, writing, and dictation.

On its own, an activity may not promote the deep thinking that leads to understanding of concepts, but when the activity is included as part of an ongoing inquiry children have time to build on what they already know to construct new knowledge and ideas, revisit their first ideas, and reflect. With ongoing inquiry activities you don't have to prep a new set of materials each week but can add just a few items to keep the

---

[42] This column entry was originally published in *Science and Children* in February 2011.

children engaged with a new challenge. They will learn about asking questions and documenting their work. Using the science and engineering practices identified in *A Framework for K–12 Science Education* (NRC 2012), a foundation for the *Next Generation Science Standards* (NGSS Lead States 2013), supports the development of science inquiry.

An in-depth science inquiry is an ongoing investigation in which children are introduced to materials through hands-on experiences and, with teacher guidance, begin to investigate a question that they can answer through their own actions and observations and with teacher-assisted research. Qualities that make an experience appropriate to include in early childhood science inquiry are described as being interesting to children, linked to their experiences, and accessible to all children's direct exploration. As children pose questions and seek answers, teachers can use these qualities to decide whether the question has the potential for ongoing inquiry. Although not all proposed activities are embraced by all my students, some of the activities that are most likely to lead into ongoing inquiry are focusing on how water moves using various kitchen tools and tubes during water play, exploring inclined planes with ramps and balls, and observing plant growth and weather-related changes in a life cycle.

When a child shows interest, use productive questions to focus attention on observations and problems to be tackled, such as, *What do you notice about…?* and *Can you find a way to…?* Your interest and questions will support their investigation. Guide the exploration by offering appropriate materials, reading books related to the topic, and initiating conversation and discussion that promote reflection on children's actions and thinking.

As teachers we grow as we learn more, so we shouldn't be embarrassed to be beginners. Experience science inquiry yourself, increase the amount of science content you integrate in your daily routine, and try introducing a topic of interest for inquiry in your classroom.

References

National Research Council (NRC). 2012. A *framework for K–12 science education: Practices, crosscutting concepts, and core ideas.* Washington, DC: National Academies Press. *www. nap.edu/catalog/13165/a-framework-for-k-12-science-education-practices-crosscutting-concepts.*

NGSS Lead States. 2013. *Next Generation Science Standards: For states, by states.* Washington, DC: National Academies Press. *www.nextgenscience.org/next-generation-science-standards.*

Additional Resources

Chalufour, I., and K. Worth. 2006. Science in kindergarten. In *K today: Teaching and learning in the kindergarten year,* ed. D. F. Gullo, 95–106. Washington, DC: National Association for the Education of Young Children. Also available online as Reading #56: Science in kindergarten. A reading from the CD accompanying *Developmentally appropriate practice in early childhood programs serving children from birth through age 8,* 3rd ed., eds. C. Copple and S. Bredekamp. Washington, DC: NAEYC, 2009. *www.rbaeyc.org/resources/Science_ Article.pdf.*

Harlen, W., and A. Qualter. 2009. *The teaching of science in primary schools.* 5th ed. London: David Fulton.

Moriarty, R. F. 2002. Entries from a staff developer's journal … helping teachers develop as facilitators of three- to five-year-olds' science inquiry. *Young Children* 57 (5): 20–24.

Activity: Properties of Water

Objective

To engage children in ongoing inquiry by exploring a question or material (in this case, the properties of water) for a period of weeks, using productive questions to stimulate further exploration and discussion to develop understanding of the concepts

Materials

- Water

- A variety of tools, such as pipettes, droppers, spoons, cups, funnels, turkey basters, sponges, small colanders, and tubes

- A variety of materials, such as felt, paper towel, plastic wrap, wax paper, wooden block, and a raincoat

- Water table or tabletop plastic tub

- Drawing and writing materials

- Camera (optional)

- Resource books on early childhood science inquiry (see "Resources" section for a few examples)

Procedure

1. Ask yourself: What are your children interested in? Observe them closely to see how they express interest though actions (what they look at and do) and questions they voice.

2. List activities that have potential for engaging the children in a long-term inquiry because they are interesting to children, linked to their experiences, and accessible to all children's direct exploration. Ask yourself whether they meet the criteria for a good early childhood inquiry experience.

3. Brainstorm or research activities that might extend or tie into the initial activity. For an activity on properties of water, keeping in mind that water can be liquid, solid, or gaseous, begin with exploring how a drop is formed and how it moves or is absorbed on paper. Explore the circumstances in which change from one state of water to another happens due to changes in location and weather.

4. Prepare productive questions and additional materials to promote further questioning and problem solving: *Do water drops move the same way on all surfaces?* (Use other materials.) *How tiny a drop can you make with a spoon or with a pipette?* (Introduce various size droppers and pipettes, big and small spoons, and spoons with holes.) *I wonder if a bowl of water will change if we put it outside today? Will a cup of hot water also steam the way the cup of coffee steamed in the air today? Where do you think an ice cube will melt fastest?* (In cold weather a variety of setups can be designed to observe changes in state.)

5. Provide materials for children to draw and write about what they see and think, to record comparisons and measurements, and to list questions they would like to answer. If you have a camera, document the children's work with photographs.

6. Hold group meetings and individual conversations to discuss the ongoing work and prompt further thinking with productive questions and new materials related to the work. Ask, *Do big drops behave the same way as small drops?* and *What happens when the ice cube warms up?* Investigate what happens when water is put into a freezer, and talk about ice on local natural bodies of water. Refer to the children's drawings and your photographs and have the children talk about what they were thinking as they worked. Ask, *And what do you think now?* or *Have your ideas grown or changed?*

Extension

A walking field trip can be built into the weekly schedule before or after recess, so children can observe the same area over a school year as they learn about the seasonal changes in animal and plant life. Journaling about an inquiry is a way to support children's continued questioning. The inquiry you and your children choose will be as individual as your class is.

Resources

Burnett, R. 1999. *The pillbug project: A guide to investigation.* Arlington, VA: National Science Teachers Association.

Chalufour, I., and K. Worth. 2003–2005. Young Scientist Series. [Individual titles are *Exploring Nature With Young Children* (2003), *Building Structures With Young Children* (2004), and *Exploring Water With Young Children* (2005).] St. Paul, MN: Redleaf Press.

Worth, K., and S. Grollman. 2003. *Worms, shadows, and whirlpools: Science in the early childhood classroom.* Portsmouth, NH: Heinemann.

Zan, B., and R. Geiken. 2010. Ramps and pathways: Developmentally appropriate, intellectually rigorous, and fun physical science. *Young Children* 65 (1): 12–17. Also available online at *www.naeyc.org/files/yc/file/201001/ZanWeb0110.pdf.*

Teacher's Picks

"Promoting Children's Science Inquiry and Learning Through Water Investigations" by Cindy Hoisington, Ingrid Chalufour, Jeff Winokur, and Nancy Clark-Chiarelli, in *Young Children* 69 (4): 72–79, 2014; available at *www.naeyc.org/yc/article/Promoting_Childrens_Science_Inquiry_Hoisington.*

This article describes how preschool teachers were supported by a professional development program, Foundations of Science Literacy (designed, implemented, and evaluated by educators and researchers at Education Development Center, Inc.), as they implemented a long-term investigation of water.

Water by Ellen Lawrence (Bearport, 2013).

Each chapter is an investigable question with a detailed list of materials needed. The author provides some answers at the end of the book.

"Let's Get Messy! Exploring Sensory and Art Activities With Infants and Toddlers" by Trudi Schwarz and Julia Luckenbill, in *Young Children* 67 (4): 26–34, 2012.

Inquiry into the properties of matter can begin in toddler classrooms—this article has tips for infant and toddler explorations.

All The Water in the World by George Ella Lyon, illustrated by Katherine Tillotson (Atheneum Books for Young Readers, 2011).

The poem and illustrations show how water is present on Earth and affects all living organisms. The water cycle is alluded to but not described scientifically.

"Teachers on Teaching: What Happens When a Child Plays at the Sensory Table" by Debra Hunter, in *Young Children* 63 (5): 77–79, 2008.

Take the author's suggestion and use the guiding questions to closely observe children's play at a sensory table to uncover children's learning in many domains.

"Why ... Take the Lid Off the Water Table?" by Cindy Gennarelli and Hope D'Avino Jennings, in *TYC (Teaching Young Children)* 1 (3): 24, 2008.

Read this one-page article to learn why exploration at the water table supports the development of children's skills and get tips on how to get started.

A Drop of Water by Gordon Morrison (Houghton Mifflin, 2006).

A read-aloud book that is also good to view up close as the author takes readers from a close-up of a single drop of water to a bird's-eye view of the landscape water flows over.

A Drop of Water: A Book of Science and Wonder by Walter Wick (Scholastic, 1997).

Use the close-up photographs showing water in motion in liquid and solid forms to get your students talking about their experiences with water.

Related *Early Years* Blog Posts

The following information is from "Sensory Table Explorations of Matter," published December 19, 2013:

> *Experiences are the beginning of understanding science and engineering concepts. When 2-year-olds explore materials and make a mixture, they are learning about the properties of "matter." Water is a favorite kind of matter for many children and adults. In warm weather, early childhood programs often make large amounts of water available for children to work with, and most children don't mind being wet. In colder weather, when getting wet means having to change clothes, adults prefer that children work with smaller amounts of liquid water or frozen water.*

Read more at *http://nstacommunities.org/blog/2013/12/19/sensory-table-explorations-of-matter.*

Additional related blog post:

"Supporting Children's Interests," published November 22, 2014: *http://nstacommunities.org/blog/2014/11/22/supporting-childrens-interests.*

Chapter 42

Sharing Research Results[43]

BEFORE YOU BEGIN THIS ACTIVITY:

Remember that each activity is not a stand-alone science or engineering curriculum. Activities are small steps in a journey of science inquiry, as discussed in Chapter 1. Your students will learn more about this concept and about the nature of science if you use this activity as part of an ongoing exploration of a question, a concept, or a topic being investigated by your class. Ask yourself, "What should come before this and what should come after?" Refer to Table 1.1 (pp. 15–39) to find other activities from The Early Years column that address the same concepts.

There are many ways to share a collection of data and your students' thinking about the data. Your students can collect data about an activity happening in your classroom, such as an exploration of which cardboard bridge design is the sturdiest for toy cars, whether the size of a seed determines how fast it sprouts, or the qualities of various playdoughs made in school or purchased. Young children can do apple-tasting tests, chart their preferences, and then share their findings with their families and other classes. Collected drawings and writings about the hatching of a caterpillar and its growth to adult butterfly form can become a book to send home or display in the school library. Collecting, analyzing, and interpreting data and communicating information are part of the science and engineering practices described in *A Framework for K–12 Science Education* (NRC 2012). Explaining the results of

---

[43] This column entry was originally published in *Science and Children* in April 2011.

science inquiry is important—that is how the body of scientific knowledge is added to by working scientists as well as amateurs.

With adult guidance, children can make playdough by following a procedure (recipe) every month in an early childhood classroom. Each time, a different recipe can be tested and the product used and evaluated. This information can be recorded in journals, pictures, and charts to be shared. What are the qualities your students desire in a playdough? Stiff or soft, bland or strong-smelling, quick-drying or never-hardening, textured or smooth? There may be many opinions about which is the "best" dough. Through preparing playdough in school, children learn about the properties of the ingredients, measurement, how to compare attributes, and how to record data.

Conversation about how the ingredients feel and whether they are wet or dry or something else will tell you what your students know already about the different states of matter, while developing their vocabulary. For primary students, vocabulary words can be added to a word wall for future reference.

Which measuring tool is used depends on the amount and the nature of the ingredient. Tools that measure the exact amount needed are easiest for young children to understand. I once made the mistake of using a 2-cup measuring cup when the recipe called for 1 cup—this confused the children, who wondered why we were only using "half" of the measuring cup. Some recipes can be made using "units" that are nonstandard measuring tools, such as baby formula scoops or small drinking cups. Write the procedure (recipe) in text and with drawings or photographs of the actual ingredient packaging and the measuring tools so even nonreaders can follow along. By making two batches at the same time, more children can participate and the groups can compare the resulting dough to see whether they are different, perhaps indicating a difference in following the procedure (recipe).

Save a sample of each completed batch in a plastic bag to compare with a sample of other playdoughs that you make or buy. Children can compare the samples and list their observations and descriptions on a chart. Ask your students, *Can any of the observations be measured with tools?* How will the children let others know which dough was the smoothest, which one was the most enjoyable to use (a highly subjective observation), which one crumbled the easiest, and which one smelled the best (also subjective)? This question is designed to encourage young children to think about how to measure the attributes rather than expecting them to understand the distinction between objective and subjective observations. Discussion can lead to developing possible tests for comparing the playdoughs and making a chart for data collection or a book of descriptive narration.

Reference

National Research Council (NRC). 2012. *A framework for K–12 science education: Practices, crosscutting concepts, and core ideas.* Washington, DC: National Academies Press. *www. nap.edu/catalog/13165/a-framework-for-k-12-science-education-practices-crosscutting-concepts.*

Activity: Playdough Guidebook

Objective
To record and report on data about various playdoughs to compare their characteristics (e.g., texture, smell, stiffness, and ability to harden)

Materials

- Playdough recipe drawn and written for children's ease of understanding

- Poster-size paper

- Ingredients for playdough recipe (e.g., flour, cornstarch, cornmeal, oatmeal, salt, cream of tartar, alum, oil, apple sauce, cinnamon, and water)

- Measuring and mixing tools such as cups and scoops, graduated measuring cups, teaspoons and tablespoons, bowls, and stirring spoons

- Safety goggles

- Drawing and writing materials

- Commercially available playdoughs (optional)

- Camera (optional)

Teacher Preparation
Make a poster of the recipes the class will be following to make playdough. You can use photos of the actual ingredients. Choose two or more recipes that will be significantly different in some way.

Procedure
CAUTION: Check for student allergies to all ingredients. Everyone should wear safety goggles to protect the eyes because young children are exuberant stirrers. (Children have fun wearing real science safety equipment.)

1. Introduce the idea of mixing materials together to make playdough, and display the ingredients in the original packaging.

2. Have the children match the packages to the pictures in the recipe and identify the measuring tools to be used for each ingredient.

3. Follow the steps of the recipe, and if cooking, complete this step of the recipe as a teacher demonstration. *CAUTION: Cook the dough by yourself if children cannot be kept away from the hot stove.*

4. Have each child feel a small amount of each ingredient at a time while discussing the properties of the materials (e.g., wet, dry, greasy, cold). A designated writer who does not have messy hands can record these observations.

5. Use the measuring steps to review number recognition and the concepts of less and more.

6. Have the children play and discover with the dough, and offer challenges such as making a ball to test how well the dough holds together and making a "snake" to test how well the dough can be shaped—its malleability. Students can measure and record the longest snake that can be made with the dough. Ask the children what they think of the dough they made and record comments.

7. Make two or three different doughs for students to test, compare, and describe on a data chart. Ask which characteristics of the doughs can be compared: for example, texture (smooth, soft, stiff, crumbly, or sticky), wetness, and smell. Children communicate the results of their investigation by recording or dictating their observations and measurements on the data chart, using descriptive words and both relative measurements and quantitative measurements (e.g., "very crumbly" and "can make a snake 24 cm long without breaking." (See Figure 42.1 for an example of a playdough data chart; the chart is also available at *www.nsta.org/elementaryschool/ connections/201104PlayDoughDataChart.pdf.*) Children can also indicate on the chart whether they want to make the playdough again.

8. Ask students to review each recipe to think about which ingredients may have contributed to the special characteristics of the dough.

The students' assessments of the various playdoughs can be combined with the recipes into a useful Playdough Guidebook to share with other classes and their families.

Resources

Playdoughrecipe.org. *www.playdoughrecipe.org/category/playdough-recipes.*

Teachnet.com. n.d. Play dough recipes: Do-it-yourself options for kids. *http://teachnet.com/ lessonplans/art/play-dough-options-do-it-yourself-recipes-for-kids.*

FIGURE 42.1. Example of playdough data chart

Record the observations on the data chart by listing descriptive words and relative measurements such as "very crumbly" and more quantitative measurements such as "can make a snake 24 cm long without breaking." Children communicate the results of their investigation by filling in the column entries under "Length of longest snake," "Texture," "Wetness," and "Smell" and by circling "Yes" or "No" under "We want to make this dough again."

Name of playdough	Length of longest snake	Texture					Wetness			Smell				We want to make this dough again.
		Smooth	Soft	Stiff	Crumbly	Sticky	Dry	A little wet	Too wet	Strong	Faint	Pleasant	Icky	
Traditional	Yes/No	Yes/No	Yes/No	Yes/No	Yes/No	Yes/No	Yes/No	Yes/No	Yes/No	Yes/No	Yes/No	Yes/No	Yes/No	Yes/No
Commercial	Yes/No	Yes/No	Yes/No	Yes/No	Yes/No	Yes/No	Yes/No	Yes/No	Yes/No	Yes/No	Yes/No	Yes/No	Yes/No	Yes/No
Cornstarch	Yes/No	Yes/No	Yes/No	Yes/No	Yes/No	Yes/No	Yes/No	Yes/No	Yes/No	Yes/No	Yes/No	Yes/No	Yes/No	Yes/No

Teacher's Picks

Pretend Soup and Other Real Recipes: A Cookbook for Preschoolers and Up by Mollie Katzen and Ann L. Henderson (Tricycle Press, 1994).

This book has recipes for food that appeals to children. The recipes are in both text and pictures.

Mudworks: Creative Clay, Dough and Modeling Experiences by MaryAnn F. Kohl (Bright Ideas for Learning; Bright Ring, 1989).

Recipes for playdough are divided into "Cooked" and "Uncooked" sections. Look for recipes for playdough and other mixtures that have symbols indicating that they can air dry, can be baked for permanence, and are edible; note that some recipes have symbols indicating that caution is suggested or that adult supervision is required.

The Piggy in the Puddle by Charlotte Pomerantz, illustrated by James Marshall (Aladdin, 1989).

Read this book with children who don't like to get their hands messy, to provide an example of characters overcoming initial reluctance to touching a messy mixture.

Pancakes for Breakfast by Tomie dePaola (HMH Books for Young Readers, 1978).

Children can "read" the pictures in this wordless story about planning to make pancakes, finding a recipe, and gathering the tools and ingredients (some from animals on the farm). Pictures show the process of making butter from fresh cream and refer to getting syrup from maple trees. Problems occur but are overcome by devising alternative plans.

Related *Early Years* Blog Posts

The following information is from "Making Playdough Is Science," published December 13, 2009:

> *What science skills will children learn while making playdough? How can making a material for play support developing math skills and language and literacy development? Some examples of skills learned while making playdough include recognizing that symbols represent real things, measuring, following the steps of a procedure in the right order, mixing materials to transform them, using imagination, and artistic expression.*
>
> *Playdough is easy to make with children because there is some "wiggle room" in the amounts—a little more water will make a softer dough, a little less oil will make it a bit sticky. Recipes for playdough (a valuable classroom tool) are widely available online and in activity books.*

Read more at *http://nstacommunities.org/blog/2009/12/13/making-playdough-is-science*.

Additional related blog post:

"Sharing Research Results of Play Dough Comparison," published April 10, 2011: *http://nstacommunities.org/blog/2011/04/10/sharing-research-results-of-play-dough-comparison*.

Chapter 43

Measuring Learning[44]

BEFORE YOU BEGIN THIS ACTIVITY:

Remember that each activity is not a stand-alone science or engineering curriculum. Activities are small steps in a journey of science inquiry, as discussed in Chapter 1. Your students will learn more about this concept and about the nature of science if you use this activity as part of an ongoing exploration of a question, a concept, or a topic being investigated by your class. Ask yourself, "What should come before this and what should come after?" Refer to Table 1.1 (pp. 15–39) to find other activities from The Early Years column that address the same concepts.

As teachers, we assess children's learning to understand how to make our instruction more effective, but assessment of children in early childhood classrooms "is challenging because they develop and learn in ways that are characteristically uneven and embedded within the specific cultural and linguistic contexts in which they live" (Copple and Bredekamp 2009, p. 22). During individual and group discussions, children often reveal their understanding of concepts. For example, I assessed what a group of 4-year-olds knew about measurement as we talked about how much water had filled a rain gauge. Using mathematics (measuring) is part of the science and engineering practices described in the *Next Generation Science Standards* (NGSS Lead States 2013).

An understanding of measurement develops in early childhood with experience. Being able to understand a "one-to-one" correspondence (saying one number for each item as we count, or matching one to one) is a skill we hope children

---

[44] This column entry was originally published in *Science and Children* in July 2011.

will develop before kindergarten. Learning how to measure and count are skills needed to collect data and recognize patterns in the data, which are part of the science and engineering practice Using Mathematics and Computational Thinking, described in *A Framework for K–12 Science Education* (NRC 2012). To teach these skills, teachers must assess how much the students already know and what they learn, as described in the National Association for the Education of Young Children 2003 position statement on early childhood curriculum, assessment, and program evaluation.

As we looked at the column of water inside a rain gauge, I asked the class of 4-year-olds, "How can we describe this amount of water to other people?" The children pointed to the numbers (inches of rain) marked on the column, recognizing them as measurement even though they were not familiar with the decimals indicating tenths of an inch. They also selected unit cubes or measuring hands (a length of paper handprints taped together) and links of chain to hold alongside the water column to see how tall it was. Other children decided to make a drawing (some with the numbers included) to show the height, while others compared the water height with an object such as a pencil or a length of beads. With each new method of measuring I had the children count or compare and then asked whether anyone had another measuring method that they would like to try.

This activity allowed me to assess what the children understood about measurement. One child did not point to each link while counting but chanted numbers when moving a finger along the length of a chain. Another child measured with a chain and then drew a fairly accurate picture of the height of the water column, adding the numbers bunched together at the bottom. By assessing where children are in their understanding of measurement, I can plan activities that are within each student's capabilities while teaching the next step.

References

Copple, C., and S. Bredekamp, eds. 2009. *Developmentally appropriate practice in early childhood programs serving children from birth through age 8.* 3rd ed. Washington, DC: National Association for the Education of Young Children.

National Association for the Education of Young Children (NAEYC) and the National Association of Early Childhood Specialists in State Departments of Education (NAECS/SDE). 2003. *Early childhood curriculum, assessment, and program evaluation: Building an effective, accountable system in programs for children birth through age 8.* Washington, DC: NAEYC and NAECS/SDE. *www.naeyc.org/files/naeyc/file/positions/pscape.pdf.*

National Research Council. 2012 (NRC). A *framework for K–12 science education: Practices, crosscutting concepts, and core ideas.* Washington, DC: National Academies Press. *www.nap.edu/catalog/13165/a-framework-for-k-12-science-education-practices-crosscutting-concepts.*

NGSS Lead States. 2013. *Next Generation Science Standards: For states, by states.* Washington, DC: National Academies Press. *www.nextgenscience.org/next-generation-science-standards.*

Activity: Temperature Changes

Objective
To determine what children know about temperature changes due to solar radiation absorption

Materials

- Small plastic bags

- Tape

- Ice cubes

- Thermometer (optional; temperature changes are more apparent on larger thermometers so use one that is ~45 cm tall, if possible)

- Permanent marker

- Drawing and writing materials

Procedure

1. Before going outdoors with the class, comment on whether you can feel warmth from the "Sun's energy" on your skin (on sunny or cloudy days).

2. Lead a group discussion about warmth felt on the skin in sunlight, how strong the effect is, and what else sunlight could warm up in addition to our skin. During discussion and the following investigation, note what children already know about sunlight and shadow formation by recording their comments and actions. (As an extension, shadow play indoors with flashlights or an overhead projector as a light source can help children learn how shadows form.)

3. Outside, the class can walk around touching various objects (tree trunk, grass, blacktop, swing, monkey bars, school building) both in the sunlight and in shadow, testing for "warm" or "cool." *CAUTION: Be aware of any surfaces that get hot enough to cause burns, and do not let children touch them.*

4. Tell the children to plan a way to show whether sunlight can change the temperature of objects. Ask, *What could we do to find out?* and *How would*

we know—how could we measure whether the sunlight is warming an object? Children may suggest feeling objects and recording data—the relative temperature as felt by individual students. Each child can record sites in order of temperature, coldest to warmest, with drawings on paper. Another method is to put the same size ice cubes in small plastic bags and tape them to various objects and in the sunlight or in shadow. Discussion about where to put the bags will show whether the children understand that shadows form where objects block the sunlight. A child who puts the bag "in the soft grass so it won't melt" does not understand the relationship between the sunlight and heat. Older children who read numerals may want to measure using a thermometer.

5. On a bright sunny day (for greatest difference in results) have each child tape a plastic bag holding an ice cube to an object outdoors, some in sunlight and some in shadow.

6. Have the children predict whether their ice cube will melt more or less, and record this with a permanent marker on the bag or on a chart.

7. Draw a chart to record the location and relative amount of melting at set intervals (5–15 minutes depending on the air temperature).

8. At the set intervals have the children investigate all the bags to see how melted the ice cubes are: "a little," "some," "a lot," or "all melted." Each child can record the amount of melting of their cube on a large group chart (see Figure 43.1 or the NSTA Connections chart at *www.nsta.org/ elementaryschool/connections/201107EarlyYearsSampleGroupChart.pdf* for an example). By comparing all the ice cubes, the children get a larger set of data than if they just looked at their own and are better able to judge the relative amount of melting.

Disagreements may arise over the amount of melting and lead to discussion—an important part of science investigations. The students may decide they need to more precisely measure the degree of melting—a refinement that can happen the next time they do this activity. They may have additional questions such as, "Will an ice cube in a bag melt on a 0° day in any location? And will it stay frozen for more than a few minutes on a 100°F day in any location?"

FIGURE 43.1. Example of data collection chart for melting ice cubes

Location	5 minutes	10 minutes	15 minutes	20 minutes	25 minutes
In sunlight					
On top of slide—Umar	Melted a little	Melted a lot			
On grass—Mitch	Not melted	Melted some			
On picnic table—Rose	Not melted	Melted some			
In shadow					
Under sliding board—K'nesha	Not melted	Not melted			
On grass in shadow—Gerson	Not melted	Melted a little			
Under picnic table—Christina	Not melted	Not melted			

Teacher's Picks

Thinking BIG, Learning BIG: Connecting Science, Math, Literacy, and Language in Early Childhood by Marie Faust Evitt (Gryphon House, 2009).

Chapter 7, "Thinking BIG About Ice: Brrr, It's Cold!" has many explorations about cold and frozen water, including relative measurement by sense of touch and standard measurement with a thermometer.

Temperature: Heating Up and Cooling Down by Darlene Stille, illustrated by Sheree Boyd (Amazing Science; Picture Window Books, 2004).

An introduction to "temperature" and "heat" helps teachers and children think about how heat moves and is measured.

Hot and Cold by Angela Royston (My World of Science; Heinemann Library, 2001).

The pictures and text are a good introduction to the vocabulary of sensing and measuring temperature. The book includes some explorations that are appropriate for early childhood programs.

On the Same Day in March: A Tour of the World's Weather by Marilyn Singer, illustrated by Frané Lessac (Harper Children's, 2000).

This book takes readers to locations around the world, some of which will be hotter or colder than the reader's. Children can infer the relative temperature by looking at the types of clothing people wear.

Related *Early Years* Blog Posts

The following information is from "Heat and Energy: What Can Young Children Understand?" published February 9, 2015:

> As a preschool teacher I am not confined to teaching concepts and vocabulary during a "unit" but can engage children in discussions and revisit activities throughout the year. In summer, we might feel heat radiating from a metal sliding board or the blacktop. The metal and blacktop were warmed as they absorbed the radiation from the Sun. Year-round, we can experience the changing temperature of a cup of hot (warm) chocolate, a baked potato, or a hard-boiled egg cooling down in our hands as the heat transfers to our hands and the surrounding air. Another way of exploring the transfer of heat is to melt an ice cube in our hands.

Read more at *http://nstacommunities.org/blog/2015/02/09/heat-and-energy-what-can-young-children-understand*.

Additional related blog posts:

"Concepts That Cut Across Science Disciplines," published January 26, 2014: *http://nstacommunities.org/blog/2014/01/26/concepts-that-cut-across-science-disciplines*.

"More Snow? Counting and Science in Winter Cold," published March 2, 2014: *http://nstacommunities.org/blog/2014/03/02/more-snow-counting-and-science-in-winter-cold*.

Chapter 44

A Sense of Place[45]

BEFORE YOU BEGIN THIS ACTIVITY:

Remember that each activity is not a stand-alone science or engineering curriculum. Activities are small steps in a journey of science inquiry, as discussed in Chapter 1. Your students will learn more about this concept and about the nature of science if you use this activity as part of an ongoing exploration of a question, a concept, or a topic being investigated by your class. Ask yourself, "What should come before this and what should come after?" Refer to Table 1.1 (pp. 15–39) to find other activities from The Early Years column that address the same concepts.

When looking at a globe, children often ask, "Where are we?" showing that they understand the globe represents a place. But their descriptions of travel show their incomplete understanding of distance and time or relative position of places. They may say, "I visited my grandparents who live in [a neighboring state]. It's over there, about three minutes," pointing in the wrong direction. Children's ability to read and create maps—to develop a sense of place—grows with experience (Sobel 1998). Understanding the geography of their community can begin with expanding their spatial relationship knowledge of their school grounds.

Children who recognize toys that are models and build models of objects themselves are ready to consider neighborhood maps as representations of the actual surroundings. Using models to explain one's thinking is one of the science and engineering practices and involves two of the crosscutting concepts described in *A Framework for K–12 Science Education* (NRC 2012): Scale, Proportion, and Quantity and Systems and System Models.

---

[45] This column entry was originally published in *Science and Children* in September 2011.

Take a walking field trip to look for familiar sights while helping children build mental and visual maps of their area. The walk can become a scavenger hunt as children look for particular objects such as street signs, buildings, and plants. These objects connect to other areas of curriculum—literacy, geometry, and biology. Street signs are some of the first words and symbols children learn to "read," even before they understand the relationship of letters to meaning. Photographs (taken by the teacher ahead of time) of expected sights can be matched to the actual objects on the walk. When children discover the object shown in the photo they are carrying, they can place the photo onto a teacher-made map in the order that it is encountered.

Mapping is a difficult task for most children younger than age 4. Using other kinds of models in the classroom teaches children the concept of representing real objects (e.g., sticks to build a "house" or a banana as a telephone). Children can understand that models have limitations. When I introduced a class of 4-year-olds to plush toy bird models, one child remarked that the beak of the toy robin was longer than the beak of the toy chickadee. We began comparing the models and listening to the recording inside the toys. Although the chickadee and robin beaks were different lengths, the bodies were the same size and shape, an inaccuracy that we checked by looking at a book with life-size drawings. We then discussed how models are not exact replicas of real things—the doors of toy cars don't open, teddy bears don't have claws or fur, and the toy birds did not have feet. The children accepted that the toy birds were useful models for the general shape and colors, the shape of the beak, and the accurate song (recordings of actual birds). The toys were obviously not alive, something you may need to mention with very young children.

Talk with your class about the use and limitations of models to prepare the children to understand maps as simplified models of a larger area. Then take them on a walking field trip to create a model of the area they traverse, broadening their understanding of the usefulness of models and their school's place in the landscape.

References

National Research Council (NRC). 2012. *A framework for K–12 science education: Practices, crosscutting concepts, and core ideas.* Washington, DC: National Academies Press. *www. nap.edu/catalog/13165/a-framework-for-k-12-science-education-practices-crosscutting-concepts.*

Sobel, D. 1998. *Mapmaking with children: Sense of place education for the elementary years.* Portsmouth, NH: Heinemann.

Activity: Schoolyard as Model

Objective

To become familiar with the purpose of maps as tools to serve as models of a larger area

Materials

- Trash bag

- Poster-size map of field trip route (to be made by the teacher)

- Photos of objects of interest along route (to be made by the teacher; lamination or enclosing in plastic reclosable sandwich bags is optional)

- Tape

- Drawing materials, identification books or handouts for animals and plants, magnifiers, small clear containers with lids for holding small animals, snack, camera (all of these are optional items)

Teacher Preparation

Walk the field trip route and take photographs of the objects you want the children to focus on—street signs, land forms, particular plants, or structures. Print the photos as individual cards for the children to carry (laminate or bag them individually so each child can carry one photo) and make a poster-size map of the route including some landmarks such as the school building, notable trees, and streets. *CAUTION: Check the field trip area for dangerous trash and natural hazards such as poison ivy, mushrooms, rough terrain, or water, so you can warn the group ahead of time to avoid these hazards. Follow school guidelines for safety.*

Procedure

1. Read aloud a book about maps (see, e.g., the books in the "Resources" section). In a group discussion, show the walking field trip map and help children decode your symbols for the landmarks. Trace the route you will be walking and ask the children what they expect to see. *CAUTION: Remind the children of safety rules appropriate to your route. For example, the teacher is the line leader, stay with your adults, look for the object in your photograph but don't pick it up, and stay on the sidewalk.*

2. Show the children the process of attaching a photo to the poster map using tape. Make the bathrooms the last stop before walking out of the building.

3. Give each child a photo to hold to search for the pictured object.

4. Stop and take a look around frequently as you wait for stragglers to catch up (with an adult at the back). Ask the children to tell you what they see, and talk about what you see. Recite poetry or sing a song as a group to keep everyone focused.

5. As the children find their pictured objects, help them attach the photos to the map in the appropriate place. Allow them time to think about where to put the photos in relation to the starting point or landmarks. Differences in spatial knowledge will be revealed as some children struggle to make sense of the map while others predict what is coming up.

6. At the end of the field trip, after all the photos are attached, briefly review the route and ask children to recall their observations.

If a map seems beyond your children's understanding, create a similar route in miniature, using blocks on a large sheet of paper to represent the landscape. Have the children "walk" dolls along the route, comparing it to what they see outdoors. With additional walking field trips along the same route, the children will develop their understanding of the spatial relationships of the objects shown on the map. Have children draw their own maps of the classroom, the playground, or the community. Challenge map-wise children to draw a map of a familiar story.

Resources

Hartman, G. 1993. *As the crow flies: A first book of maps*. New York: Aladdin Books.

Leedy, L. 2003. *Mapping Penny's world*. New York: Henry Holt.

Teacher's Picks

"A Walk in the Woods" by Cynthia Hoisington, Nancy Sableski, and Imelda DeCosta, in *Science and Children* 48 (2): 27–31, 2010.

This article thoroughly describes how a preschool class can benefit from a field trip.

Window by Jeannie Baker (Walker Books, 2002).

The perspectives shown on the front cover (bird's eye view) and the inside pages (window view) encourage readers to think about what they can see from different vantage points.

Down the Road by Alice Schertle, illustrated by E. B. Lewis (Harcourt, 2000).

The maplike illustration on a front page sets the stage for changing points of view as a child goes down the road to the store, for the first time by herself. Children can recall the views and landmarks they see on their way to school and look forward to a time when they can walk somewhere by themselves.

Mouse Views: What the Class Pet Saw by Bruce McMillan (Holiday House, 1994).

Your children will see the classroom in a new way after seeing how a mouse might view it.

Related *Early Years* Blog Posts

The following information is from "Walking Fieldtrips to Draw Nature," published September 17, 2008:

> *A walking field trip can bring much-needed outdoor time and opportunity for scientific observation to a class schedule. The objective can be to view the sky, look for birds, find seeds, or inventory the surrounding environment. Whether it's just around the school building or to a natural area several blocks away, a walking field trip is most successful when materials for dealing with unexpected discoveries are brought along. These materials may include identification books or handouts for birds, trees, and insects; magnifiers; small containers for holding small animals; a camera; and drawing and writing materials. Make sure children dress appropriately, and bring snacks and emergency supplies.*

Read more at *http://nstacommunities.org/blog/2008/09/17/walking-fieldtrips-to-draw-nature.*

Additional related blog posts:

"Models, and Maps, and Spatial Understanding," published September 9, 2011: *http://nstacommunities.org/blog/2011/09/09/models-and-maps-and-spatial-understanding.*

"Preschool Experiences in a Winter Forest," published March 15, 2015: *http://nstacommunities.org/blog/2015/03/15/preschool-experiences-in-a-winter-forest.*

"Spring Wildflowers: Introducing Guest Blogger Marie Faust Evitt," published April 28, 2012: *http://nstacommunities.org/blog/2012/04/28/spring-wildflowers-introducing-guest-blogger-marie-faust-evitt.*

Chapter 45

Reading Stories and Making Predictions[46]

BEFORE YOU BEGIN THIS ACTIVITY:

Remember that each activity is not a stand-alone science or engineering curriculum. Activities are small steps in a journey of science inquiry, as discussed in Chapter 1. Your students will learn more about this concept and about the nature of science if you use this activity as part of an ongoing exploration of a question, a concept, or a topic being investigated by your class. Ask yourself, "What should come before this and what should come after?" Refer to Table 1.1 (pp. 15–39) to find other activities from The Early Years *column that address the same concepts.*

I like to read picture books to my students as part of science learning, choosing books that include science concepts, either directly or indirectly. The text often has rich vocabulary, we can talk about what could happen and what is pretend, and we learn as much from the illustrations as we do from the text. Books present many opportunities for children to make claims (also called predictions) about what might happen next based on the evidence in the story and to reflect on them—goals that are part of most English and science education standards. *A Framework for K–12 Science Education* (*Framework*; NRC 2012) describes the importance of being able to predict and to construct scientific explanations and solutions. Children can use their observations about events or patterns in nature as evidence for their ideas about causes. In the *Next Generation Science Standards,* Cause and Effect: Mechanism and Explanation is one of the crosscutting concepts and Engaging in Argument From Evidence is one of the science and engineering practices (NGSS Lead States 2013).

---

[46] This column entry was originally published in *Science and Children* in November 2011.

National Science Teachers Association

"What might happen if … ?" Children ask this question often, through actions rather than words when they are very young, and verbally as they mature. For example, at 2 years old, children like to explore the concepts of floating and sinking but don't pause to think about what might happen before they toss an object into the water. With practice planning (e.g., what area of the room they are going to work in or what block they will put on top next), children begin to think about what comes first, next, and last in time and in space. They begin to sort objects based on an attribute they choose. When exploring buoyancy, they test their choice by putting objects into the water, one by one. At some point children are able to predict. Planning gives them time to practice and develop their self-regulation, an important developmental goal for early childhood (NAEYC 2009), as well as time to formulate and ask questions (e.g., "Why do some things sink and some things float?").

Teachers who ask productive questions (Elsteeg 1985), without stopping children's exploration by giving a correct answer, allow children to examine an object or problem closely, measure and count, compare, classify, and find out "what happens if." Predicting can begin with talking about what comes next in the daily classroom routine or what the weather might be like at recess time. When reading and listening to literature, children can use evidence from the pages that came before to predict what will come next. Make it clear that you want the children to make the predictions and that you will respect and accept all answers but will not provide any.

Your class may want to write its own book—maybe about a typical day in school. The children's artwork or photography can illustrate the book, and children who don't yet write can dictate longer text. To encourage readers to predict, children should include explicit questions such as, "What do you think happened then?"

References

Elsteeg, J. 1985. The right question at the right time. In *Primary science: Taking the plunge,* ed. W. Harlen, 36–46. Oxford, UK: Heinemann.

National Association for the Education of Young Children (NAEYC). 2009. *Position statement: Developmentally appropriate practice in early childhood programs serving children from birth through age 8. www.naeyc.org/files/naeyc/file/positions/PSDAP.pdf.*

National Research Council (NRC). 2012. *A framework for K–12 science education: Practices, crosscutting concepts, and core ideas.* Washington, DC: National Academies Press. *www. nap.edu/catalog/13165/a-framework-for-k-12-science-education-practices-crosscutting-concepts.*

NGSS Lead States. 2013. Appendix F—Science and engineering practices in the *Next Generation Science Standards,* and]Appendix G—Crosscutting concepts. In *Next Generation Science Standards: For states, by states.* Washington, DC: National Academies Press. *www.nextgenscience.org/next-generation-science-standards.*

Activity: "What Might Happen If … ?"

Objective

To have students think about a story and make predictions about what might happen next

Material

A fiction book to read aloud that is age-appropriate for your class

Teacher Preparation

Ask your school or local librarian for help finding a picture book or chapter book (depending on the age of the students) that has some connection to the science concept that you are focusing on (see "Resources" section for examples). If none are available, use a book that has a simple storyline in which students can guess what will happen next at every turn of the page.

Procedure

1. Read the selected book when there is ample time for student discussion. Tell your students that you are going to read a book aloud and that you will stop now and then to ask them to tell what they think will happen next in the story—that is, to make a prediction. Remind students that scientists don't always agree with each other and that you predict that *all ideas will be received with respectful listening.*

2. Model making a prediction by looking at the cover of the book and think aloud about what you think the book will be about, and why you think so.

3. Read a few pages of the book and ask, *What do you think might happen next?* or *What do you think [the main character] should do?* It might be appropriate to ask these questions at every page of a picture book, but not as often with a chapter book.

4. To encourage all students to participate, have students draw a picture of what they think will happen next or, more simply, a picture of one object that they think will be on the next page.

5. Ask students to tell why they think a particular scenario will happen next, and discuss the ideas students expressed with words and drawings.

6. At the end of the book ask students whether the story turned out the way they predicted it would. Have students write, dictate, or draw a different ending.

Extension

Your students can share the fun of predicting what comes next in a story by creating their own book to share with family members or another class. Their drawing and writing abilities will determine the length of individually made books. A class book can have one page by each student following a story line or story starter that you suggest or that is decided on by the class. Dictation can help students express complex thoughts, and shared writing keeps students involved as much as possible. I wonder what your students will write about?

Resources

Charlip, R. 1964. *Fortunately.* New York: Aladdin.

McNaughton, C. 1998. *Suddenly! A Preston pig story.* Orlando, FL: Harcourt, Brace.

TeacherVision. n.d. Teaching strategies for reading. *www.teachervision.fen.com/reading/ resource/48646.html.*

Teacher's Picks

The Magic Hat by Mem Fox, illustrated by Tricia Tusa (Harcourt, 2002).

The improbable wind and a magic hat create a storyline that allows children to make many predictions.

"When a Hypothesis Is NOT an Educated Guess" by Louise M. Baxter and Martha J. Kurtz, in *Science and Children* 38 (7): 18–20, 2001.

Bark, George by Jules Feiffer (HarperCollins, 1999).

After reading the first few pages in which the puppy, George, makes the sounds of other animals, children will want to predict what sound he'll make next and whether he will ever bark like a dog.

Suddenly! by Colin McNaughton (HMH Books for Young Readers, 1998).

A bit scary but with a happy ending, this book pairs a possible action with what actually happens. Invite children to predict what Preston the pig will do next, or have them say what they would do.

The Wind Blew by Pat Hutchins (Aladdin, 1993).

As the wind captures objects from townspeople, children can predict where the objects will end up.

Related *Early Years* Blog Posts

The following information is from "Sink? Float? Try It With Pumpkins," published October 22, 2008:

> *For an activity to explore buoyancy—what materials and which objects sink or float in water—I gave each child in a small group an object to hold. Then I explained that we were going to think about the objects and say where we think they will come to rest in a big tub of water—at the bottom or near the top of the water—BEFORE we put the objects in the water. Most of the 2-year-olds are in the "thought is action" stage and immediately dropped (threw, in some cases) the objects in the tub. "I think it's going to … It's floating!"*

Read more at *http://nstacommunities.org/blog/2008/10/22/sink-float-try-it-with-pumpkins*.

Additional related blog posts:

"Playing With Magnets and Learning About the Property of Materials," published December 7, 2008: *http://nstacommunities.org/blog/2008/12/07/playing-with-magnets-and-learning-about-the-property-of-materials*.

"What Will Happen If I . … Preparing an Effective Learning Environment With Help From Peep," published December 5, 2014: *http://nstacommunities.org/blog/2014/12/05/what-will-happen-if-i-preparing-an-effective-learning-environment-with-help-from-peep*.

Chapter 46

Seeing the Moon[47]

BEFORE YOU BEGIN THIS ACTIVITY:

Remember that each activity is not a stand-alone science or engineering curriculum. Activities are small steps in a journey of science inquiry, as discussed in Chapter 1. Your students will learn more about this concept and about the nature of science if you use this activity as part of an ongoing exploration of a question, a concept, or a topic being investigated by your class. Ask yourself, "What should come before this and what should come after?" Refer to Table 1.1 (pp. 15–39) to find other activities from The Early Years column that address the same concepts.

Spotting the Moon in the sky is like finding a treasure—unexpected and beautiful. When children look for the Moon in the sky, they don't know where to look. This is true of many adults, too. A few months of daily "Moon observation journaling" will reveal the pattern of changes in appearance and position and lead to understanding that there is a relationship—the Moon does not appear at random times.

If we look closer with binoculars or a telescope or view a high-quality photograph or video, we see that the Moon's body is round like a sphere and the surface is covered with many circular shapes. These circles are called *impact craters* (NASA 1997a, 1997b). The craters form when pieces of another body in space, such as an asteroid or comet, hit the surface of a planet or a natural satellite (such as the Moon). Craters on the Moon are easy to see because, unlike on Earth, there is no weather to make changes to the surface.

The Moon is far away and most easily observed at a time when most young children are sleeping.

---

[47] This column entry was originally published in *Science and Children* in January 2012.

Because direct contact isn't possible, adults have to be creative in how they help children learn about the Moon. Even a single night of Moon viewing guided by an interested adult can support the development of young children's curiosity and their image of themselves as explorers of the natural world beyond Earth. I remember when my father took me and my sisters outside (later than our bedtime) to see the Moon using an old telescope. The telescope view made the Moon appear to be within reach, so bright and detailed, and suddenly so real.

We can help young children begin to understand one of the reasons for the patterns we see on the Moon's surface by modeling the process that creates impact craters. Understanding about cause and effect is one of the crosscutting concepts in *A Framework for K–12 Science Education* (NRC 2012), which recognizes that in the earliest grades, children "begin to look for and analyze patterns … [and] begin to consider what might be causing these patterns and relationships" (p. 88). Developing and using models is one of the science and engineering practices and is also a grade 1 performance expectation (1-ESS1 Earth's Place in the Universe, which includes observations of the Moon to describe patterns that can be predicted) in the *Next Generation Science Standards* (NGSS Lead States 2013).

Making impact craters in sand by dropping several balls of different sizes or weights can be a simple activity for students to learn how craterlike shapes are formed in the sand. It can be expanded into an activity in which relative measurements are made, comparing the size of the ball, the size of the resulting crater, the height from which the ball is dropped, and the depth of the crater. This chapter describes a simple crater formation activity, but you can add measurement depending on the readiness of your class.

Support children's developing language arts skills by holding discussions for oral language practice and introducing new vocabulary—*crater, circle, sphere*, and *impact*. Have children document the process of crater formation with drawings or photographs so they will look closely and think about the pulling force (of gravity) that brought the dropped balls into the sand and the pushing force of the ball that moved the sand into the crater shape. For younger children, this activity teaches about using models to understand Earth science processes. Older children may also begin to think about force by comparing craters made by different weight balls and the push the children give them.

References

National Aeronautics and Space Administration (NASA). 1997a. Lesson 6: Impact craters—Holes in the ground! In *Exploring meteorite mysteries: A teacher's guide for activities with Earth and space sciences*. EG-1997-08-104-HQ. *http://solarsystem.nasa.gov/docs/Impact_Craters_Holes_508FC1.pdf.*

National Aeronautics and Space Administration (NASA). 1997b. Teacher Page: Impact craters. In *Exploring the Moon—A teacher's guide with activities*. EG-1997-10-116-HQ. *www.nasa.gov/pdf/180572main_ETM.Impact.Craters.pdf*.

National Research Council (NRC). 2012. *A framework for K–12 science education: Practices, crosscutting concepts, and core ideas*. Washington, DC: National Academies Press. *www.nap.edu/catalog/13165/a-framework-for-k-12-science-education-practices-crosscutting-concepts*.

NGSS Lead States. 2013. *Next Generation Science Standards: For states, by states*. Washington, DC: National Academies Press. *www.nextgenscience.org/next-generation-science-standards*.

Activity: Crater Making

This activity was adapted from *Thinking BIG, Learning BIG: Connecting Science, Math, Literacy, and Language in Early Childhood* by M. F. Evitt, T. Dobbins, and B. Weesen-Baer (Gryphon House, 2009).

Objective
To explore the formation of impact craters

Materials

- Photograph of the Moon (*Note:* A high-resolution photo of a full Moon taken from the Apollo 11 spacecraft on July 19, 2013, is available at *www.nasa.gov/mission_pages/apollo/40th/images/apollo_image_25.html*.)

- Damp, not wet, sand—outside or in a tub indoors—or dry baking soda, enough for a depth of 10 cm

- Safety goggles

- Objects to drop into the sand—marbles, ball bearings, and balls of varying weights: foam ball, table tennis ball, golf ball, and baseball (include two balls of the same size but different weights, if possible)

- Tongue depressor or piece of stiff cardboard

- Standard or nonstandard measurement tools (e.g., centimeter ruler or linked paper clips)

- If indoors, a tablecloth under the tub to make cleanup easier (and prevent slips)

Teacher Preparation

Prepare the outside area or indoor tub of sand or baking soda. Test the activity beforehand to be sure the sand or baking soda will hold an impression of a ball dropped into it.

Procedure

1. As a group, view a photograph of the Moon and discuss its appearance. If possible, view a high-resolution photograph on a large screen so details are easily seen.

2. Take notes of the children's ideas and descriptions of their previous experiences and knowledge about the Moon. Children may say, "I see circles!" "What made the craters?" "Why is part of the Moon light-colored and part of it dark?"

3. Ask the children, *Can you think of a way that we can make craters in some material?* Help the children plan the setup, including the use of safety goggles, by asking open-ended questions such as, *What could we do to make big craters and small craters?* and *How can we protect our eyes from any sand that gets thrown into the air?*

4. Have the children handle the balls and weights and predict what kind of impact crater each one might make. Emphasize the importance of simply dropping the balls from the same height (without throwing or pushing them down) to form craters, so that the size and weight of the balls will be the only different factor to affect the size and depth of the craters. (After the first tries you can add the force of the throw or drop as a variable.)

5. Have the children stand and drop a ball from shoulder height into the sand. *CAUTION: Have all participants, children and teachers, put on safety goggles.* At first the children will all want a turn and will want to make the biggest, deepest crater. After their first tries they will become more interested in discovering how to make craters of different sizes and depths.

6. After each impact, have children remove the object and smooth the sand surface with a tongue depressor or piece of stiff cardboard to prepare for the next impact.

7. Challenge the children to measure the crater sizes and depths, and document how they were made by making a chart or drawing pictures and dictating the procedure. Ask, *Is there a way you can make a smaller/larger or deeper/more shallow crater?*

Extension

The children may become interested in how impressions of all kinds are made in various materials. Continue this exploration using modeling clay, potter's clay, and other dough or clay types, comparing the force needed to push and shape different materials.

Teacher's Picks

Next Time You See the Moon by Emily Morgan (NSTA Kids, 2014).

The beautiful photos will help children wonder about the phases of the Moon. Exploring shadow formation is a foundational experience for later understanding about objects in our solar system.

Crab Moon by Ruth Horowitz, illustrated by Kate Kiesler (Candlewick Press, 2000).

Children will read about a nighttime adventure that was aided by an occurrence of the full Moon.

Related *Early Years* Blog Posts

The following information is from "Moon and the Earth and the Sun, and More," published January 6, 2012:

> *Craters on the Moon are visible without a telescope, even in daytime, when teachers can point out the Moon to their students. See the Madison Metropolitan School District's Planetarium website at* https:// planetarium.madison.k12.wi.us/mooncal/daymoon *to learn more about viewing the Moon in daytime. Children enjoy making craters in snow, damp sand, or other fine material, by dropping balls of varying sizes and weights into the material.*

Read more at *http://nstacommunities.org/blog/2012/01/06/moon-and-the-earth-and-the-sun-and-more.*

Additional related blog posts:

"Early Learning Experiences Build Toward Understanding Concepts That Are Hard to Teach," published September 27, 2012: *http://nstacommunities.org/blog/2012/09/27/early-learning-experiences-build-toward-understanding-concepts-that-are-hard-to-teach.*

"More Science in the Early Years—A Reoccurring Theme From High School Teachers and Researchers," published January 28, 2010: *http://nstacommunities.org/blog/2010/01/28/more-science-in-the-early-years-a-reoccurring-theme-from-high-school-teachers-and-researchers.*

Chapter 47

Circle Time[48]

BEFORE YOU BEGIN THIS ACTIVITY:

Remember that each activity is not a stand-alone science or engineering curriculum. Activities are small steps in a journey of science inquiry, as discussed in Chapter 1. Your students will learn more about this concept and about the nature of science if you use this activity as part of an ongoing exploration of a question, a concept, or a topic being investigated by your class. Ask yourself, "What should come before this and what should come after?" Refer to Table 1.1 (pp. 15–39) to find other activities from *The Early Years* column that address the same concepts.

The first few weeks of school are a time for establishing classroom routines. Introducing regularly scheduled science-centered discussions in the form of a circle time is a worthwhile addition, reinforcing communication and listening skills in the context of science exploration. "Science talks" have been regarded as a way to encourage "children to describe what they have seen and done and present their evidence and thoughts ... raise questions about one another's evidence and ideas, and connect their own work to that of others" (Worth and Grollman 2003). Science talks are also a time for children to use two of the science and engineering practices: Engaging in Argument From Evidence and Obtaining, Evaluating, and Communicating Information (NRC 2012).

Hold regular, short discussions to build children's capacity to participate and contribute, and lengthen them as children become used to routines. Begin with 10-minute circle time, and if children are actively involved, gradually expand it to a developmentally appropriate 20 minutes for 4-year-olds (Copple and Bredekamp 2009). Ask "productive" questions about the presented evidence to encourage children to think about an answer from what they know rather than find the answer in your question (Elsteeg 1985). If teachers use the same question structure too often, children may be able to guess the correct answer without thinking. In a science talk, productive questions might be, *Tell me what you noticed about the object. What happens if you … ?* or *Why do you think that?*

---

[48] This column entry was originally published in *Science and Children* in July 2012.

Explicitly teach and reinforce your expectations to allow cooperative, productive work and discussion (Wilson 2011). With routines established early, students can focus on gaining experience and learning content. Some of the routines I use to make a science talk go more smoothly include using a song or chant for passing materials around the group (e.g., "And pass it to your neighbor") and using silent signals if I am reading aloud or in discussion (e.g., hands-on-the-head signal that they can't see the book).

We can teach the class how to listen and ask questions by modeling and describing the expected behavior (Wilson 2011). Don't be discouraged if the science circle discussion is not a total success the first time. Revisit the procedure with changes that aim to redirect children's behavior to successfully talk with each other. The following activity uses a "teach-and-tell" format to introduce children to speaking with their classmates about what they observe or know.

References

Copple, C., and S. Bredekamp, eds. 2009. *Developmentally appropriate practice in early childhood programs serving children from birth through age 8.* 3rd ed. Washington, DC: National Association for the Education of Young Children.

Elsteeg, J. 1985. The right question at the right time. In *Primary science: Taking the plunge,* ed. W. Harlen, 36–46. Oxford, UK: Heinemann.

National Research Council (NRC). 2012. *A framework for K–12 science education: Practices, crosscutting concepts, and core ideas.* Washington, DC: National Academies Press. *www. nap.edu/catalog/13165/a-framework-for-k-12-science-education-practices-crosscutting-concepts.*

Wilson, M. B. 2011. *What every kindergarten teacher needs to know.* Turner Falls, MA: Northeast Foundation for Children.

Worth, K., and S. Grollman. 2003. *Worms, shadows, and whirlpools: Science in the early childhood classroom.* Portsmouth, NH: Heinemann.

Activity: Teach and Tell

Objective

To develop a routine for class discussion about science topics that will support children in contributing to the group, listening to each other, interacting cooperatively, learning from each other, and asking questions

Materials

- Simple nonfiction book (see step 2)

- Student's natural object or a drawing, photograph, or description of an object or event (one object or drawing, photograph, or description for each child)

- Large pad of paper or whiteboard for drawing

Teacher Preparation

Keep a box of small intriguing natural objects in storage for children who forget to bring in an object. You can add items that relate to current science explorations.

Procedure

1. In the first week of school, tell the children that they each will have a time to show the rest of the class a natural object that they like and talk about it during a "Teach and Tell" science circle (or any name that fits your class) for a small group or the entire class. Send the same information home in a note to families explaining the reasoning and the restrictions, or include it in your first newsletter (see sample note below).

   Dear families,

   This year students will share something from nature that interests them with the rest of the class during our "Teach and Tell" time. It may not be a toy or a human-made object. The object should be something they found, or a drawing or photograph of something they found, such as a leaf, a drawing of the sky, or the family pet; or it may be an event, such as "It rained." They may tell something about the object or event or share a question they have about it. The rest of the class will tell their ideas about the object or event and add their questions about it to the discussion. I will send home a note when it will be your child's turn, with a list of suggestions. Your child may also choose something from the school yard or classroom to talk about.

2. For the first Teach and Tell science circle, read a simple nonfiction book (find one in the school or public library [see, e.g., the Rookie Read-About Science series published by Children's Press], or search for appropriate titles at NSTA Recommends: *www.nsta.org/recommends*). Before reading, tell the class that afterward they can take turns telling the group about something they learned from the book, sharing something they already

know if it has to do with the book, and asking questions. After reading the book, model how to share an idea by saying a few sentences or drawing on a large pad or board. Then ask the children to tell or draw their ideas about the book and ask questions. Record their words on paper or a whiteboard, using collaborative writing if children are ready, and save the notes to refer to in later discussions.

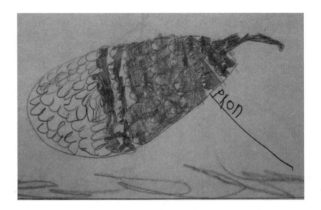

3. As the weeks progress, observe the children closely and change the science circle structure as needed. Have the presenting student begin to call on those who have questions and help record their statements, if possible. If children have a topic or object similar to a previous one, you can say, *Scientists often repeat the work of other scientists to find out if they have the same observations.* If children disagree with another's statements, you can say, *Scientists don't always agree. It's okay to have different ideas. We can say "I agree" or "I disagree" and tell why.* A statement or drawing that some disagree with can have a question mark put next to it, to show that further exploration is needed (Sander and Nelson 2009). The note can be removed if exploration or research resolves the question.

Later in the year, the Teach and Tell science circles can expand into a time to share evidence and discuss the collected data from any science investigations that have been developed in your class. Children may then be able to talk directly to each other rather than waiting to be called on by the teacher or a lead student and ready to record some of their own words. By establishing a Teach and Tell science circle in the first weeks of school, you will have a routine that will facilitate communication and science learning and help children become comfortable speaking in a group.

Reference

Sander, J., and S. Nelson. 2009. Science conversations for young learners: Tips on guiding kindergarteners to participate in large-group discussions in science. *Science and Children* 46 (6): 43–45.

Resources

Chalufour, I., and K. Worth. 2006. Science in kindergarten. In *K today: Teaching and learning in the kindergarten year*, ed. D. F. Gullo, 95–106. Washington, DC: National Association for the Education of Young Children (NAEYC). Also available as *Reading #56, Science in kindergarten*. A reading from the CD accompanying *Developmentally appropriate practice in early childhood programs serving children from birth through age 8*, 3rd ed., eds. C. Copple and S. Bredekamp. Washington, DC: NAEYC, 2009. *www.rbaeyc.org/resources/Science_Article.pdf*.

Copple, C., and S. Bredekamp, ed. 2009. *Developmentally appropriate practice in early childhood programs serving children from birth through age 8*. 3d ed. Washington, DC: National Association for the Education of Young Children.

Copple, C., S. Bredekamp, and J. Gonzalez-Mena. 2011. Q&A with the editors of Developmentally Appropriate Practice. *www.naeyc.org/event/developmentally-appropriate-practice?page=1*.

Gartrell, D. 2006. Guidance matters: The beauty of class meetings. *Young Children: Beyond the Journal*, November. *www.naeyc.org/files/yc/file/200611/BTJGuidance.pdf*.

Teacher's Picks

"Talk Strategies: How to Promote Oral Language Development Through Science" by Lauren M. Shea and Therese B. Shanahan, in *Science and Children* 49 (3): 62–66, 2011.

The authors describe talk strategies that provide many opportunities for students to speak and teachers to check for understanding.

"Whatever Happened to Developmentally Appropriate Practice in Early Literacy?" by Susan B. Neuman and Kathleen Roskos, in *Young Children: Beyond the Journal*, July 2005, available at *www.naeyc.org/files/naeyc/file/Publications/ArticleExamples/02Neuman.pdf*.

Read about developmentally appropriate, content-rich literacy learning strategies to implement with your science teaching.

Backyard Detective: Critters Up Close by Nic Bishop (Scholastic 2003).

Just one page of this book of photo collages of life-size backyard environments will inspire questions for a group discussion.

Maple Trees by Allan Fowler (Rookie Read-About Science; Children's Press 2001).

Reading aloud a book that shows a natural environment is a good way to begin using a Teach and Tell science circle strategy. Look for books that reflect your local environment.

Related *Early Years* Blog Posts

The following information is from "Teach and Tell Circle Time," published June 25, 2012:

> *Some programs have a "Teach and Tell" circle time that has several purposes—to provide a focused time to learn about natural materials, to allow children to each have a turn as the circle leader by talking and taking questions, to learn how to make a question, and to practice group discussion skills. Only natural objects are shared, because when the sharing child brings a toy, the focus turns to each child wanting to tell about their own toys rather than ask questions. Drawings or photographs of natural objects are also allowed. If a child brought in a wooden toy to share we could have a discussion about whether or not it should be permitted—a good way to help teach the difference between "natural" and "manufactured."*

Read more at *http://nstacommunities.org/blog/2012/06/25/teach-and-tell-circle-time.*

Additional related blog post:

"Beginning the Year With a Plan to Support Science Talk," published July 5, 2012: *http://nstacommunities.org/blog/2012/07/05/beginning-the-year-with-a-plan-to-support-science-talk-through-circle-time-discussions.*

Chapter 48

Drawing Movement[49]

BEFORE YOU BEGIN THIS ACTIVITY:

Remember that each activity is not a stand-alone science or engineering curriculum. Activities are small steps in a journey of science inquiry, as discussed in Chapter 1. Your students will learn more about this concept and about the nature of science if you use this activity as part of an ongoing exploration of a question, a concept, or a topic being investigated by your class. Ask yourself, "What should come before this and what should come after?" Refer to Table 1.1 (pp. 15–39) to find other activities from The Early Years column that address the same concepts.

When introducing the materials for exploring the motion of objects, I often ask children to tell me how to make a ball move on a level ramp. They usually want to demonstrate with a push of their hand, or by lifting one end of the ramp, but I stop them and repeat that I want them to "use their words" to tell me what their idea is.

They may or may not have the vocabulary to express what they think—something that becomes apparent when they have to talk about their ideas rather than show them with gestures. I wait for the child to use whatever vocabulary she or he has before providing some words to fill out the description. A child might say, "I go [gestures with a hand pushing motion]," and I respond, "You *push* the ball to make it move on the ramp?" This conversation is both assessment and a teaching moment for learning vocabulary.

Rolling balls down ramps is a way to explore the physical science concept of motion and the engineering practices of Planning and Carrying Out Investigations and Designing Solutions (NGSS Lead States

---

[49] This column entry was originally published in *Science and Children* in November 2012.

2013). Part of the kindergarten performance expectation K-PS2, Motion and Stability: Forces and Interactions, is to "plan and conduct an investigation to compare the effects of different strengths or different directions of pushes and pulls on the motion of an object" (NGSS Lead States 2013).

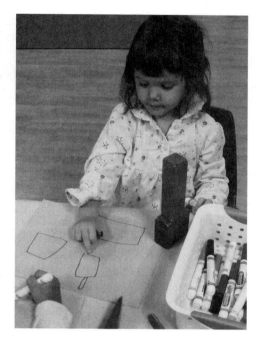

Children are fascinated by watching balls or marbles move quickly down ramps, but they rarely take time to stop and reflect on what happened. They may not notice that this time they pushed the marble, rather than just letting go. They may not pay attention to how the slope of the ramp affects the speed of the marble. Making a drawing of what happened on their ramp structure is a way to both assess what children understand about the physical action they started and teach children about this tool for reflecting on their understanding. After children have had several weeks of open-ended exploration building and using the ramp structures, I ask them to draw what they have built as a way to explain what happens to objects (usually some sort of ball or marble) as they move on the structure.

When children draw the structure and represent how a ball or marble moves down a ramp, they observe the ball or marble motion more closely. They think about the action and the structures they designed as they work on their drawings. Describing their drawings to a teacher and classmates helps children think again about what happened and share their understanding. This discussion is part of teaching visual literacy—the skills needed to interpret and construct meaning from images of many kinds and the ability to create such images. The drawings are also a way to share learning with classmates and family.

Reference

NGSS Lead States. 2013. *Next Generation Science Standards: For states, by states.* Washington, DC: National Academies Press. *www.nextgenscience.org/next-generation-science-standards.*

Activity: Just Draw It—Ramps

Objective
To develop children's ability to describe what they know through drawing

Materials

- Balls, building blocks, and ramp materials (e.g., cardboard tubes, wooden cove molding trim [available at do-it-yourself building supply stores], plastic rain gutters)

- Drawing materials (*Note:* Pipe cleaners or wax sticks such as Wikki Stix or Bendaroos can also be used to make lines on paper to represent ramp structures.)

- Camera (optional)

Teacher Preparation

Prepare yourself to introduce and work with students on building ramps and exploring the motion of balls by reading about this physical science activity and by trying it yourself (see "Resources" section).

Procedure

1. Provide a variety of materials for building ramps and objects to move or roll on the ramps. Introduce the class to the idea of designing ramp structures and exploring motion of objects on the ramps.

2. Allow students to use the materials for long periods. Researchers advise allowing children to work through their problems themselves as they design, try, and redesign their structures (Stoll et al. 2012). While observing students, ask open-ended questions to support student exploration of motion.

3. As an introduction to drawing a structure, challenge students to try to draw one or two building blocks. Gently discourage children if they want to trace the block by asking them to *just draw it.* If children become frustrated, don't insist but demonstrate how to draw the simple rectangle or other shape that represents the block. A two-dimensional drawing is appropriate—perspective drawing isn't necessary.

4. After students have had four to six weekly hour-long sessions building and using ramps, have them try to make pictures of their ramp structures to explain how a ball moves on it. Pictures that do not look like the ramp structure still represent the children's idea of the motion. What is important is allowing the students to describe their work, not teaching them how to accurately draw a ramp structure. Children may use symbols they have seen regularly used in other media—animation in video and drawings in books or cartoons—including arrows and grouped lines to

represent motion of the object, or they may use multiple images of an object to represent its path along the ramp. They develop visual literacy as they work to explain motion using a two-dimensional media.

5.   During small-group or whole-class discussions about the motion of balls on the ramps, have children explain their drawings (see Figure 48.1 for examples of drawings). If a student has used symbols such as sweeping lines or drawing the object in several places along the ramp to show motion, be sure to say, *Tell me about the meaning of this part of your drawing* to get the student's observations of motion. Students can also draw objects and represent motion on photographs of structures printed on plain paper.

6.   Children may write or dictate a description of the motion pictured. As you discuss a picture with students, look for evidence in their drawing that they understand more than they can verbalize and help them build vocabulary by describing what you think you see and asking for their confirmation.

DeVries and Sales (2011, p. 7) advise teachers, "Be patient. Most children experiment with the same problems of force, motion, and the like for some time before they construct complex knowledge about the phenomena."

FIGURE 48.1. Examples of children's drawings about the motion of balls on ramps

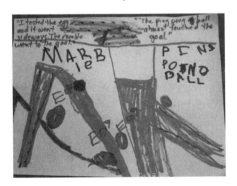 

References

DeVries, R., and C. Sales. 2011. *Ramps and pathways: A constructivist approach to physics with young children.* Washington, DC: National Association for the Education of Young Children.

Stoll, J., A. A. Hamilton, E. Oxley, A. Mitroff Eastmand, and R. Brent. 2012. Young thinkers in motion: Problem solving and physics in preschool. *Young Children* 67 (2): 20–26.

Resources

Peep and the Big Wide World. 2015. Explore ramps. *http://peepandthebigwideworld.com/en/educators/curriculum/family-child-care-educators/14/ramps.*

University of Northern Iowa, College of Education, Regents' Center for Early Developmental Education. n.d. Ramps and pathways. *www.uni.edu/rampsandpathways.*

Zan, B., and R. Geiken. 2010. Ramps and pathways: Developmentally appropriate, intellectually rigorous, and fun physical science. *Young Children* 65 (1): 12–17. Also available online at *www.naeyc.org/files/yc/file/201001/ZanWeb0110.pdf.*

Teacher's Picks

"Blocks: Great Learning Tools From Infancy Through the Primary Grades," *Young Children* **70 (1), March 2015 [entire issue].**

This issue includes two articles that address block building in preschool: "Building Bridges to Understanding in a Preschool Classroom: A Morning in the Block Center" by Lea Ann Christenson and Jenny James and "Using Blocks to Develop 21st Century Skills" by Karen Wise Lindeman and Elizabeth McKendry Anderson.

Uncovering Student Ideas in Physical Science, Volume 1: 45 New Force and Motion Assessment Probes **by Page Keeley and Rand Harrington (NSTA Press, 2010).**

The assessment probes can be used by preschool teachers to generate questions for discussion, and the probes can be used by teachers of children who read to find out what children know about motion.

Roll, Slope, and Slide: A Book About Ramps **by Michael Dahl, illustrated by Denise Shea (Amazing Science: Simple Machines; Picture Window Books, 2006).**

This book explains how ramps help people do work and includes a simple investigation.

Forces Make Things Move **by Kimberly Brubaker, illustrated by Paul Meisel (HarperCollins, 2005).**

Read parts of this book aloud, using the illustrations to review children's experiences. The text can guide discussions about concepts such as force and friction, but omit sections that go beyond the grade-level expectations.

Move It!: Motion, Forces and You by Adrienne Mason, illustrated by Claudia Davila (Primary Physical Science; Kids Can Press, 2005).

This simple introduction to motion and forces can stimulate discussion of the children's own experiences.

And Everyone Shouted, "Pull!" by Claire Llewellyn, illustrated by Simone Abel (First Look: Science; Picture Window Books, 2004).

Children can relate their own experiences to the problem solving and forces applied by farm animals as they travel by cart to market.

Building Structures with Young Children by Ingrid Chalufour and Karen Worth (Young Scientist Series; Redleaf Press, 2004).

This curriculum is a guide for teachers to intentionally help children develop understanding of physical science and engineering concepts as they play with blocks.

I Fall Down by Vicki Cobb, illustrated by Julia Gorton (HarperCollins, 2004).

The simple activities in this book can be part of an ongoing investigation into motion.

Push and Pull by Patricia J. Murphy (Rookie Read-About Science; Children's Press, 2002).

Have children identify "pushes" and "pulls" in the photographs on each page as you or they read this book.

Cubes, Cones, Cylinders, and Spheres by Tana Hoban (Greenwillow Books, 2000).

This wordless book can be used to introduce vocabulary for three-dimensional geometric shapes and to support children's thinking about how shape affects motion.

Wheel Away! by Dayle Ann Dodds, illustrated by Thacher Hurd (HarperCollins Children's, 1991).

This fun book, with a repeating phrase for children to say, also presents a question: Can a rolling object go down a hill and up another hill and then return to the starting point?

Related *Early Years* Blog Posts

The following information is from "Building With Blocks: Exploring Stability and Change in Systems," published May 10, 2015:

> *A common and simple system that children work with in early childhood programs is the structure they build from blocks. Teachers have a role in supporting children's work in block building, and professional organizations support teachers—the National Association for the Education of Young Children and the National Science Teachers Association.*

Read more at *http://nstacommunities.org/blog/2015/05/10/building-with-blocks-exploring-stability-and-change-in-systems*.

Additional related blog posts:

"Spatial Thinking," published March 24, 2011: *http://nstacommunities.org/blog/2011/03/24/spatial-thinking*.

"Learning About Shapes, With Tips From a Special Education Teacher," published January 8, 2014: *http://nstacommunities.org/blog/2014/01/08/learning-about-shapes-with-tips-from-a-special-education-teacher*.

Chapter 49

Please Touch Museum[50]

BEFORE YOU BEGIN THIS ACTIVITY:

Remember that each activity is not a stand-alone science or engineering curriculum. Activities are small steps in a journey of science inquiry, as discussed in Chapter 1. Your students will learn more about this concept and about the nature of science if you use this activity as part of an ongoing exploration of a question, a concept, or a topic being investigated by your class. Ask yourself, "What should come before this and what should come after?" Refer to Table 1.1 (pp. 15–39) to find other activities from The Early Years column that address the same concepts.

The exploration of materials happens daily as children play on natural and artificial surfaces and use various materials to create artwork and structures. This is evident when you observe your students as they search for the drawing tool that works for their purpose, intentionally decide between Legos and Magna-Tiles to build a spaceship, or choose the wet-enough snow to make snowballs. Through these types of choices children display their knowledge of the properties of materials. Young children need to experience the varied materials in the world to be able to distinguish between natural and artificial materials and to experience their properties.

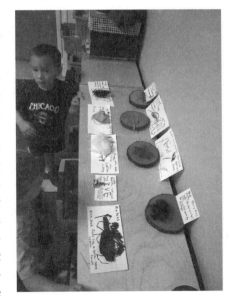

Early childhood education standards and national science standards describe the importance of learning about the properties of natural and human-manufactured materials through direct exploration. The *Head Start Early Learning Outcomes Framework: Ages Birth to Five* affirms that children's "inclination to be curious, explore, experiment, ask questions, and develop their own theories about the world makes science an important domain for enhancing learning and school success" (U.S. DHHS, ACF, Office

---

[50] This column entry was originally published in *Science and Children* in December 2012.

of Head Start 2015). Foundational experiences in describing and classifying different kinds of materials by their observable properties will build children's understanding toward the *Next Generation Science Standards* grade 2 performance expectation 2-PS1, Matter and Its Interactions: "Plan and conduct an investigation to describe and classify different kinds of materials by their observable properties" (NGSS Lead States 2013).

The National Association for the Education of Young Children's standards and criteria for accreditation recognize children's interest in collections of natural objects, as well as the importance of access to an outdoor environment that accommodates "exploration of the natural environment, including a variety of natural and manufactured surfaces, and areas with natural materials such as nonpoisonous plants, shrubs and trees" (NAEYC 2014, p. 65). Your class might want to make either a material "please touch museum" or a collection of materials for making creations in a supply box. A museum might have one of each type of natural and artificial materials regularly found at school. Objects or samples of materials can be brought in from home with permission. Children can engage their literacy skills by writing or dictating a card for each museum object describing what it is made of, evidence of its characteristics and properties, and whether it is natural or manufactured. Include objects of unknown materials to show that science exploration is ongoing. These cards can also be made for materials on the classroom supply shelves.

Product designer Inna Alesina says, "Materials are like words. The richer your design vocabulary, the more solutions you can see and express" (Alesina and Lupton 2010, p. 4). The activity seems very simple, but the conversations and design work that develop with it provide insight into what children know about materials.

References

Alesina, I., and E. Lupton. 2010. *Exploring materials: Creative design for everyday objects*. New York: Princeton Architectural Press.

National Association for the Education of Young Children (NAEYC). 2014. *NAEYC early childhood program standards and accreditation criteria and guidance for assessment*. Washington, DC: NAEYC. *www.naeyc.org/files/academy/file/AllCriteriaDocument.pdf*.

NGSS Lead States. 2013. *Next Generation Science Standards: For states, by states. www.nextgenscience.org/next-generation-science-standards*.

U.S. Department of Health and Human Services (DHHS), Administration for Children and Families (ACF), Office of Head Start. 2015. *Head Start early learning outcomes framework ages birth to five.* Washington, DC: U.S. DHHS, ACF, Office of Head Start. *http://eclkc.ohs.acf.hhs.gov/hslc/hs/sr/approach/pdf/ohs-framework.pdf.*

Activity: Materials Museum

Objective
To experience and describe a wide range of materials and discuss their attributes and uses

Materials

- A display area (if making a "please touch museum") or a storage box for supplies

- Various small classroom and school yard objects and sample-size pieces of natural or artificial, human-manufactured materials (size depends on space available for display)

- Note cards (or other small paper) and writing tools

Teacher Preparation
In conversations as children play indoors and outdoors throughout the year, describe the materials in use. For example, *These smooth, wooden blocks are made from trees, just like the sticks you gathered yesterday.* Ask open-ended questions as children work with materials to elicit their close examination and descriptions of materials, such as, *What do you notice about the fabric of that scarf?* (Use the word *fabric* instead of *material* when referring to cloth). Everyday experiences with natural and artificial materials may seem too ordinary to talk about, but with teacher guidance children will use these experiences to build understanding about materials.

Procedure

1. Gather a variety of objects and sample-size pieces made from different materials for children to handle—aluminum foil, keys, metal and plastic lids, many types of fabrics, wax paper, beeswax and candles, yarn and wire, pieces of natural and altered wood, seed pods, small tiles, stones of all kinds, pieces of art foam, and cardboard and all kinds of paper. Use fewer materials for children younger than age 4—perhaps just wood, plastic, stone, and metal. As children handle the materials, have them record on note cards or dictate their descriptions of the color, texture,

and any other attributes such as stickiness or being able to bend without breaking and their evidence of whether it is natural or manufactured. Handling and talking about materials can build vocabulary for all children, including dual-language learners (Nemeth 2014). Encourage children to talk about uses for the material. Minimally processed objects manufactured from natural materials, such as a wooden spoon, can challenge children to go beyond describing the material to thinking about its origin.

2. Depending on your students' abilities and interest, create a "please touch museum" display of materials or a supply box of materials for creative building. With students' help, if possible, make a display area for the samples of materials and objects and their description cards. This display can expand as children discover new objects to add (e.g., a plastic spoon from a snack) or bring samples from home (e.g., a pinecone). *CAUTION: Check all objects for sharp edges, allergens, or other safety concerns.* Children can sort displayed objects into groups by material, whether they are natural or artificial, or another attribute.

3. If you decide to put the material samples in a supply box, supplement them with additional sizes of the same materials and family-donated materials, to provide a larger supply for building and making. Encourage children to use the materials to design structures, models, toys, or sculptures from their imagination by asking, *What could this piece be used for?* and *Which of these materials would make a good structural support [or soft bed, window, writing surface, and so on]?*

Children need outdoor experiences with natural materials to become familiar with which materials occur naturally and which are made by people from those natural materials. For more information about materials and how people use and make them, read fiction and nonfiction books aloud (see "Teacher's Picks" section and search for books on materials at NSTA Recommends, *www.nsta.org/recommends*).

Reference

Nemeth, K. October 23, 2014. Fast 5 gamechangers that build language quality for DLLs! Language Castle. *www.languagecastle.com/2014/10/fast-5-gamechangers-build-language-quality-dlls*.

Resource

National Science Teachers Association (NSTA). NSTA recommends. *www.nsta.org/recommends*. (This webpage allows you to search for reviews of science-teaching books, including series on materials from Capstone Press, Heinemann Library, Kingfisher Books, and Lerner Publications.)

Teacher's Picks

Publications

I Had a Favorite Dress by Boni Ashburn, illustrated by Julia Denos (Harry N. Abrams, 2011).

The properties of the fabric of a favorite dress become evident as a skillful mother reshapes it as her daughter grows.

Amazing Materials by Sally Hewitt (Amazing Science; Crabtree, 2007).

Many different materials, and some of their properties, are discussed in this book.

Senses at the Seashore by Shelley Rotner (Millbrook Press, 2006).

The color photographs in this book show what children can see, hear, smell, touch, and taste at the beach.

Glass by Chris Oxlade (Materials; Heinemann Library, 2002) and other titles in the Materials series.

Books about glass, wool, and other materials help children understand that the objects they use are made from a wide variety of materials, each with its own properties.

Joseph Had a Little Overcoat by Simms Taback (Viking, 1999).

The properties of some materials allow them to be reshaped multiple times. Challenge children to think of examples of materials they reshaped for new purposes.

Website

AppLit. Appalachian Animal Tales: Variants of "The Three Little Pigs" *www2.ferrum.edu/applit/bibs/tales/other3pigs.htm.*

This bibliography by Tina L. Hanlon lists versions of the story "The Three Little Pigs," including both traditional and modern retellings. The tales introduce children to materials for building a house.

Related *Early Years* Blog Posts

The following information is from "First Week of Preschool for Two-Year-Olds," published September 16, 2013:

> *The first week of school is when we begin to know our students and make observations about their skills, personalities and interests. Sensory experiences engage them and introduce materials from the natural world—rough and smooth pieces of tree bark, fuzzy and smooth leaves, big and small leaves, leaves with a strong smell, dry and wet sand, sea shells and water in jars and tubs. The hard objects should be too large to choke on, and the softer ones, like leaves, need adult supervision to remind the children to keep them out of their mouths.*

Read more at *http://nstacommunities.org/blog/2013/09/16/first-week-of-preschool-for-two-year-olds*.

Additional related blog post:

"Sensory Experiences to Invoke an Environment Described in a Book," published April 7, 2013: *http://nstacommunities.org/blog/2013/04/07/sensory-experiences-to-invoke-an-environment-described-in-a-book*.

Chapter 50

The Wonders of Weather[51]

BEFORE YOU BEGIN THIS ACTIVITY:

Remember that each activity is not a stand-alone science or engineering curriculum. Activities are small steps in a journey of science inquiry, as discussed in Chapter 1. Your students will learn more about this concept and about the nature of science if you use this activity as part of an ongoing exploration of a question, a concept, or a topic being investigated by your class. Ask yourself, "What should come before this and what should come after?" Refer to Table 1.1 (pp. 15–39) to find other activities from The Early Years column that address the same concepts.

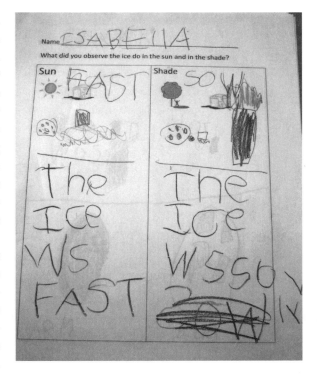

Winter weather often inspires teachers' and students' interest in collecting weather data, especially if snow falls. Beginning weather data collection in preschool will introduce children to the concepts of making regular observations of natural phenomena, recording the observations (data), and looking for patterns in data. Through this, children become familiar with making observations in a scientific investigation as an ongoing process and become more aware of the weather.

Measuring precipitation can be part of the activities your class does to build toward the *Next Generation Science Standards* kindergarten performance standard K-ESS2: Earth's Systems, as children use and share observations of local weather conditions to describe patterns over time (NGSS Lead States 2013).

---

[51] This column entry was originally published in *Science and Children* in January 2013.

A Framework for K–12 Science Education describes grade band endpoints (what students should know by the end of grades 2, 5, 8, and 12) that relate specifically to the study of weather and how climate affects living organisms (*Framework*; NRC 2012). We should keep in mind the grade 2 band endpoint for ESS2.D as students progress in learning about this disciplinary core idea: "Weather is the combination of sunlight, wind, snow or rain, and temperature in a particular region at a particular time. People measure these conditions to describe and record the weather and to notice patterns over time" (NRC 2012, p. 188).

Children use symbols and vocabulary words in recording daily weather data to represent weather events such as a cloudy day. Exposing children to print—creating a print-rich environment— supports their awareness of the importance of text (Young 2009). In comparing the recorded weather one month to the next, children count and create simple graphs of their data. Incorporating the weather data into a student newspaper "weather report" connects to social studies curriculum (Sahn and Reichel 2008). Table 50.1 lists more ideas on connecting across the curriculum.

Recording data on the temperature, amount of precipitation, presence of wind, and cloud cover to answer the question "Will the weather be the same or different as the days go by?" can become a daily routine for student engagement. If you are not living in a place with daily or seasonal weather fluctuations, another option is to make observations at different times of the day (e.g., early-morning temperature versus late-afternoon temperature). Children will enjoy having responsibility for the daily data recording. They will be practicing real science, collecting real data, comparing the data, and discussing the meaning of the differences and similarities in weekly weather.

TABLE 50.1. Examples of curriculum areas and activities connected to observing weather

Curriculum area	Activities connected to observing weather	
Literacy	Use symbols and text to represent weather events in data collection— wind, rain, cloud cover, snow.	Draw pictures and write sentences to tell about a weather event. Students share and discuss their work.
Mathematics	Measure the amount of rainfall in inches in a rain gauge, or the depth of snow. Estimate the relative amount of cloud cover—all of the sky, most of the sky, or some of the sky is covered.	At the end of each week or month, make a tally chart to count the number of days with the same weather. Make a graph comparing the numbers of different weather events that month.

Table 50.1. (*continued*)

Curriculum area	Activities connected to observing weather	
Social studies	If you think families in your program can contribute, collect gently used winter jackets, rain jackets, and boots for children in another program who might need them.	Create a student-written and -drawn weekly newspaper of significant events, including the weather.
Nature experience	Raise funds to purchase a classroom set of rain boots and jackets or umbrellas so children can walk outside during or after a rain.	Experiencing the weather is a natural experience that does not require leaving the school grounds.

Source: Adapted from NSTA Connections at *www.nsta.org/elementaryschool/connections/201301ConnectingWeather Observations.pdf.*

References

National Research Council (NRC). 2012. *A framework for K–12 science education: Practices, crosscutting concepts, and core ideas.* Washington, DC: National Academies Press. *www. nap.edu/catalog/13165/a-framework-for-k-12-science-education-practices-crosscutting-concepts.*

NGSS Lead States. 2013. *Next Generation Science Standards: For states, by states.* Washington, DC: National Academies Press. *www.nextgenscience.org/next-generation-science-standards.*

Sahn, L., and A. Reichel. 2008. Read all about it! A classroom newspaper integrates the curriculum. *Young Children* 63 (2): 12–18.

Young, J. 2009. Enhancing emergent literacy potential for young children. *Australian Journal of Language and Literacy* 32 (2): 163–180.

Activity: Observing Weather

Objective

To participate in an ongoing data collection of local weather conditions

Materials

- Weather symbol strips (see Figure 50.1)

- Rain gauge (see "Resource" section)

- Outdoor thermometer

- Temperature recording sheets (see Figure 50.2)

- Rainfall recording sheets (see Figure 50.3)

- Measuring cubes (optional—see step 1 of the "Procedure" section)

- Photographs of items representing temperature highs (e.g., a cup of liquid with steam rising) and lows (e.g., a glass of ice cubes)

- Blue and white crayons

- Quarter-sheets of blue construction paper

- Tally chart (see Figure 50.4)

- Tray (helpful for carrying materials and as a writing surface)

Teacher Preparation

Prepare materials for your class to collect weather data over a period of several months or for the school year. Print, copy, and cut out the weather symbol strips for recording individual weather observations. Make a simple rain gauge (see "Resource" section for instructions). Attach pictures representing temperature highs and lows to an outdoor thermometer for prekindergarten students to use as reference points, and set up the thermometer where students can see it. Using the templates in Figures 50.2 and 50.3, print the recording sheets for temperature and rainfall observations.

Procedure

1. Take the weather symbol strips, blue and white crayons, quarter-sheets of blue construction paper, rainfall recording sheets, and temperature recording sheets outside at recess or any other time. Model for students how to document the weather by speaking your thoughts aloud as you work. Circle the appropriate symbols for any weather conditions you observe—snowy, rainy, sunny, cloudy, or windy—on the weather symbol strips. Use a blue crayon to color in the rain gauge image to record the height of rainwater in the gauge (use nonstandard measuring units such as measuring cubes if the gauge is not marked with inch units). Color in the thermometer image to record the temperature (height of the liquid in the column). Look at the sky and draw the amount of cloud cover using a white crayon on blue construction paper. Let all children who want to practice recording their observations. The accuracy of their documentation will get better with time.

2. In the classroom, talk with students about making daily weather observations. Add "weather recorder" to the class daily job chart. One student can record weather data each day, but allow all interested students to record. Some data records can be sent home with students, but keep one record to post later.

3. Post completed daily weather data strips, cloud drawings, and temperature and rainfall recording sheets in the classroom so students can begin to compare the data.

4. After two weeks of recording weather data, have students count the number of days on which each type of weather condition occurred using a tally chart. Have students create a bar graph with counting cubes or on paper to show how many days of each weather condition occurred.

5. Have students continue documenting their weather observations, and compare the first month's tally chart of data with the next month's. To promote analysis of the data, ask questions such as, *Which weather events happened most often last month? Did any weather event happen more this month than last month?* and *I wonder if the weather will be the same next month?*

Extension

Through discussion and conversations, students will begin to see patterns in the weather data—the seasonal changes. Support their further questions by inviting expert guests to answer questions about weather, providing materials to investigate the properties of water and air, or begin a project about the changes in living organisms in the schoolyard as the season changes to spring.

Resource

Schafernaker, T. 2011. How to … build a rain gauge. BBC One. *www.youtube.com/watch?v=BdeKdT0nwow.*

Teacher's Picks

The Sunny Day by Anna Milbourne, illustrated by Elena Temporin (Usborne, 2008).

One of a series by the same author where children happily explore a day, each with a different kind of weather.

Who Likes the Wind by Etta Kaner, illustrated by Marie Lafrance (Exploring the Elements; Kids Can Press, 2006).

This book uses simple concepts of science to answer children's questions about the wind.

Millions of Snowflakes by Mary McKenna Siddals, illustrated by Elizabeth Sayles (Houghton Mifflin, 1998).

Simple rhyming text captures the joy of a winter afternoon while teaching basic counting skills.

Gilberto and the Wind by Marie Hall Ets (Picture Puffin Books; Turtleback, 1978).

Use this book as part of an investigation into wind and how its strength varies.

The Snowy Day by Ezra Jack Keats (Viking Books for Young Readers, 1962).

This classic story, winner of the Caldecott Medal, has lovely illustrations of a young boy discovering the properties of snow.

Related *Early Years* Blog Posts

The following information is from "Connecting to the Weather," published April 28, 2009:

> *I like the way the air smells just as a badly needed rain begins—it makes me think of the earth exhaling as the water soaks in (but this could be a misconception on my part). Rainfall is a significant event in children's lives, in some places a daily one and in others a rare pleasure. Rainy days usually mean that children play indoors, so they may not know how much it rained or how long. What can we do to connect children to the patterns in nature determined by precipitation?*

Read more at *http://nstacommunities.org/blog/2009/04/28/connecting-to-the-weather*.

Additional related blog post:

"Observing Weather Events," published January 3, 2013: *http://nstacommunities.org/blog/2013/01/03/observing-weather-events*.

FIGURE 50.1. Weather symbol strips for documentation

Note: Copy this figure and cut it into strips for student weather data collection. Students will *circle the phenomena they observe.*

Note: Enlarge this chart for use with the entire class.

Today's weather	Today's weather	Today's weather	Today's weather	Today's weather
Snowy	Snowy	Snowy	Snowy	Snowy
Rainy	Rainy	Rainy	Rainy	Rainy
Sunny	Sunny	Sunny	Sunny	Sunny
Cloudy	Cloudy	Cloudy	Cloudy	Cloudy
Windy	Windy	Windy	Windy	Windy

FIGURE 50.2. Temperature recording template

To represent the temperature of the air, color in on this sheet where you see the colored liquid in the glass tube on the actual thermometer—from the bulb at the bottom to the highest point of the colored liquid in the glass tube.

To represent the temperature of the air, color in on this sheet where you see the colored liquid in the glass tube on the actual thermometer—from the bulb at the bottom to the highest point of the colored liquid in the glass tube.

National Science Teachers Association

FIGURE 50.3. Rainfall gauge recording template

Color in the tube to show how much rain fell and how much is collected in the rain gauge.

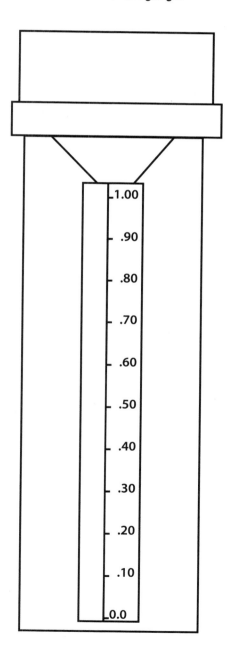

Color in the tube to show how much rain fell and how much is collected in the rain gauge.

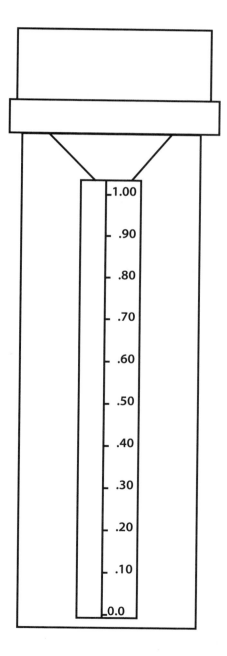

FIGURE 50.4. Tally chart for counting weekly or monthly occurrences of weather events

Note: Enlarge this chart for use with the entire class.

Snowy	Rainy	Sunny	Cloudy	Windy

Chapter 51

"Life" Science[52]

BEFORE YOU BEGIN THIS ACTIVITY:

Remember that each activity is not a stand-alone science or engineering curriculum. Activities are small steps in a journey of science inquiry, as discussed in Chapter 1. Your students will learn more about this concept and about the nature of science if you use this activity as part of an ongoing exploration of a question, a concept, or a topic being investigated by your class. Ask yourself, "What should come before this and what should come after?" Refer to Table 1.1 (pp. 15–39) to find other activities from The Early Years column that address the same concepts.

A parent of a 2½-year-old once clapped her hands over her child's ears rather than let the child hear me tell a group of children that a cricket they found was dead. "He hasn't heard that word before," she told me. The children were puzzling over why the cricket did not jump away or move. I told them that it was dead; its body did not "work" anymore. I added that all living things have life spans, and that our life spans are much longer than a cricket's. The children accepted this information matter-of-factly. They probably did not understand what death is, but they saw what it was in a cricket. Experts advise adults that death should be discussed honestly and in language that children can understand at their stage of development (see "Teacher's Picks" section).

Talking about death as part of a life cycle is often ignored or spoken about in hushed tones in early childhood. Books with "Life Cycle" in the title often do not include the death of the living organism in the information about the cycle. Rather, a new organism is born, hatches, or sprouts, and the focus shifts away from the old, parent organism to the new organism.

---

[52] This column entry was originally published in *Science and Children* in February 2013.

When a situation or a child's question brings up the subject of death, teachers should use very clear language, without substituting other words such as *sleep* or *rest* for death. We can talk with young children about how the bodies of dead animals and plants do not "work" any more. Allow children to talk about "where they went" but do not speculate on it, to keep the focus on science learning. Understanding that there is no one age at which all children are able to think about or are interested in thinking about death, teachers should respect children's varying levels of engagement with discussions and books that raise this topic (see "Teacher's Picks" section).

Early experiences observing a living organism throughout a life cycle can get children ready for a more complete understanding of life cycles by grade 3. By grade 3, the *Next Generation Science Standards* performance expectation 3-LS1: From Molecules to Organisms: Structures and Processes, includes understanding that "organisms have unique and diverse life cycles but all have in common birth, growth, reproduction, and death" (NGSS Lead States 2013). Classroom investigation of seed sprouting and growth of a plant with a short life cycle can introduce the concepts related to living organisms before beginning gardening activities.

Reference

NGSS Lead States. 2013. *Next Generation Science Standards: For states, by states.* Washington, DC: National Academies Press. *www.nextgenscience.org/next-generation-science-standards.*

Activity: A Plant's Life

Objective
To observe and document the life cycle of an organism

Materials (see also "Resources and Supplies" section, after the "Procedure" section)

- Container designed to provide continuous water supply

- Potting medium

- Light source for the container (grow light or direct sunlight)

- Seeds of the Wisconsin Fast Plant (*Brassica rapa*), a plant with a short life span

- Plan to provide the necessary care for two months

- Materials to record changes as seeds sprout and plant grows (choice of materials depends on availability)

- Centimeter ruler or nonstandard measuring tool

- Cotton swabs

- Magnifier or digital microscope (optional)

Teacher Preparation

Prepare the growing container and light source in advance. You will need one growing container that is at least 5 inches in diameter to grow four Wisconsin Fast Plants. The Wisconsin Fast Plants Program website has options for creating a continuous water supply. *CAUTION: Adults, not children, should add fertilizer when needed.* An artificial light source is necessary in most areas.

Procedure

1. Tell students that the class will plant the seeds, care for them, and observe the parts of the plants as they grow. Ask for ideas for recording the changes as the seeds grow into plants. Provide recording materials that are available in your program— paper and drawing or writing materials, digital camera, computer, printer— and decide on a system for recording every day.

2. Plant four to eight seeds. The goal is to have four plants to observe—if more than four seeds sprout, you will "thin" (remove) the weaker ones for best growth of the remaining plants. Here is a suggestion for having a class of 16–24 students plant four seeds in a way that feels "fair" to all: Have each student handle and look closely at a *Brassica rapa* seed using a magnifier or digital microscope. Each child puts "their" seed into the same cup and the teacher randomly chooses eight seeds from that cup to plant. Each child can put a pinch of planting medium on top of the planted seeds. Have students document the planting setup and process with drawings. *CAUTION: Everyone should wash their hands after handling the planting medium.*

3. Add "Plant Care and Observation" to your own calendar, put it on the daily routine, or add "Plant Caretaker" and "Plant Observer" to the job chart to ensure that someone looks at and documents any changes every day. These are indeed "fast" plants!

4. If the seeds do not sprout in four days, the Wisconsin Fast Plants Program recommends starting over. If the seedlings grow poorly or die, the class

will have to review the documentation and research to investigate what might have been the cause. Luckily, the Wisconsin Fast Plants website has information on what might go wrong and suggestions for how to fix it.

5. As the plants grow, involve your students in maintaining the water and light needs of the plants. Have students count the number of leaves, measure the height of the plant, draw or photograph the plant structure, and, when it flowers, move pollen from flower to flower with cotton swabs. As the plants grow, students will document the development of flowers and seed pods, and finally the "senescence" or aging leading to death of the plant (after 35 days).

The student documentation can be used to make a book about the life cycle of the class's *Brassica rapa* plants. Looking back at the daily observations can reveal differences between the four plants: Did all the seeds sprout on the same day? Did all the plants grow flowers at the same time? In the spring students can use their book to compare the life cycle of a *Brassica rapa* plant to the life cycles of other plants they tend in future activities, such as tulips and sugar snap peas. Noticing differences between individuals and between plant species begins students' education about genetics.

Resources and Supplies

Wisconsin Fast Plants Program: *www.fastplants.org* (home page; life cycle information is at *www.fastplants.org/life_cycle*).

Wisconsin Fast Plants seeds from Carolina Biological Supply Company: *www.carolina.com/ wfp-mutant-seed-stocks/brassica-rapa-wisconsin-fast-plants-standard-seedpack-of-50/158804.pr.*

Growing container from Bottle Biology: *www.bottlebiology.org/investigations/terraqua_build_1. html.*

Grow lights: Grow lights come in a variety of sizes. At the time of this writing, the following companies sold setups that provide light for just a few plants: Greentrees Hydroponics *www.hydroponics.net*; Growco Indoor Garden Supply *https://4hydroponics. com*; HTGSupply *www.htgsupply.com/Default.*

Teacher's Picks

The following publications and website offer advice and resources from experts about discussing death in language that children can understand at their stage of development:

Publications

Talking About Death: A Dialogue Between Parent and Child, 4th ed., by Earl A. Grollman (Beacon Press, 2011).

How Do We Tell the Children? A Step-by-Step Guide for Helping Children and Teens Cope When Someone Dies, 4th ed., by Daniel Schaefer and Christine Lyons (Newmarket Press, 2010).

"Grief: Helping Young Children Cope" by Frances B. Wood, in *Young Children* 63 (5): 28–31, 2008.

"Real Life Calls for Real Books: Literature to Help Children Cope With Family Stressors" by Sherron K. Roberts and Patricia A. Crawford, in *Young Children* 63 (5): 12–17, 2008.

Mr. Rogers' Parenting Book: Helping to Understand Your Young Child by Family Communications, Inc. (Running Press, 2002).

The following children's books raise or discuss the concept of death and how people relate to it or understand it:

The Dead Bird by Margaret Wise Brown (William Morrow, 1935/2008).

In a Nutshell by Joseph Anthony, illustrated by Cris Arbo (A Sharing Nature With Children Book; Dawn Publications, 1999).

Sophie by Mem Fox, illustrated by Aminah B. L. Robinson (Harcourt, 1989).

When a Pet Dies by Fred Rogers (G. P. Putnam's Sons, 1988).

Website

The Fred Rogers Company: Dealing With Death
www.fredrogers.org/parents/special-challenges/death.php

Related *Early Years* Blog Posts

The following information is from "Planning to Teach About Life Cycles?" published February 26, 2013:

> *There is something about the changes in the natural world due to seasonal changes in springtime that inspire us to talk about baby plants, baby birds, and baby anything. During a warm spell in January I was inspired to refurbish the garden box on the playground with some new potting soil and have all the children plant greens of various kinds by taking a pinch of the tiny seeds, sprinkling them on the soil, and patting them in. The warm days gave way to colder days, with a few below freezing, and then back up to warm. I was happily surprised to see tiny sprouts a month later, because the temperature had stayed on the cold side, mostly below 40°F. Amazing how seeds sprout when the conditions are right for growth!*

Read more at *http://nstacommunities.org/blog/2013/02/26/planning-to-teach-about-life-cycles.*

Additional related blog post:

"Early Childhood Teachers Respond to Request for Resources on Earth and Life Science," published November 15, 2011: *http://nstacommunities.org/blog/2011/11/15/early-childhood-teachers-respond-to-request-for-resources-on-earth-and-life-science.*

Chapter 52

Water Leaves "Footprints"[53]

BEFORE YOU BEGIN THIS ACTIVITY:

Remember that each activity is not a stand-alone science or engineering curriculum. Activities are small steps in a journey of science inquiry, as discussed in Chapter 1. Your students will learn more about this concept and about the nature of science if you use this activity as part of an ongoing exploration of a question, a concept, or a topic being investigated by your class. Ask yourself, "What should come before this and what should come after?" Refer to Table 1.1 (pp. 15–39) to find other activities from The Early Years *column that address the same concepts.*

One morning, after a heavy nighttime rain, the preschool sandbox crowd noticed a row of small cone-shaped holes, or pits, about 2–6 cm in diameter in the wet sand. "Hey look!" The discussion began on what could have made those holes. The children thought it might have been an animal walking or digging. "Have you ever seen footprints like these before?" I asked, trying to get them to refer to prior experiences. I noticed that the holes lined up under the edge of the awning that covers most of the sandbox. In my prior experience, water dripping off a roof makes holes in the ground below. I could have shared this experience with the children, but they could also discover it themselves, over time, through additional experiences.

Early childhood science focuses on providing experiences for young children to broaden their hands-on understanding of the natural world. Looking for and describing evidence to support ideas about natural events are part of the crosscutting concepts and the

---

[53] This column entry was originally published in *Science and Children* in April 2013.

science and engineering practices described in *A Framework for K–12 Science Education* (*Framework*; NRC 2012); these practices include Planning and Carrying Out Investigations, Constructing Explanations, and Engaging in Argument From Evidence. Supporting students in science discussions during which they share and argue about their evidence for their ideas begins in early childhood, when their experiences and any documentation and collected data are their evidence.

Teachers also introduce science concepts such as learning about how water shapes the Earth's surface to build student understanding toward the *Next Generation Science Standards (NGSS)* grade 2 performance expectation 2-ESS1-1: "Use information from several sources to provide evidence that Earth events can occur quickly or slowly" (NGSS Lead States 2013). The *Framework* and the *NGSS* begin with kindergarten, but prekindergarten teachers can look to see what lies ahead for their students or use early childhood learning standards such as the *Head Start Early Learning Outcomes Framework: Ages Birth to Five* as a guide (U.S. DHHS, ACF, Office of Head Start 2015).

By working with sand and pouring water, children can discover how water moves sand to create different shapes in sand. They can control and vary the water stream direction, amount, and speed. Being able to directly explore the materials over time, and make the changes themselves, helps children build their understanding of science concepts (Chalufour and Worth 2006).

References

Chalufour, I., and K. Worth. 2006. Science in kindergarten. In *K today: Teaching and learning in the kindergarten year,* ed. D. F. Gullo, 95–106. Washington, DC: National Association for the Education of Young Children (NAEYC). Also available online as Reading #56. Science in kindergarten. A reading from the CD accompanying *Developmentally appropriate practice in early childhood programs: Serving children from birth through age 8,* 3rd ed., eds. C. Copple and S. Bredekamp. Washington, DC: NAEYC, 2009. *www.rbaeyc.org/resources/Science_Article.pdf.*

National Research Council (NRC). 2012. A *framework for K–12 science education: Practices, crosscutting concepts, and core ideas.* Washington, DC: National Academies Press. *www.nap.edu/catalog/13165/a-framework-for-k-12-science-education-practices-crosscutting-concepts.*

NGSS Lead States. 2013. *Next Generation Science Standards: For states, by states.* Washington, DC: National Academies Press. *www.nextgenscience.org/next-generation-science-standards.*

U.S. Department of Health and Human Services (DHHS), Administration for Children and Families (ACF), Office of Head Start. 2015. *Head Start early learning outcomes framework: Ages birth to five.* Washington, DC: U.S. DHHS, ACF, Office of Head Start. *http://eclkc.ohs.acf.hhs.gov/hslc/hs/sr/approach/pdf/ohs-framework.pdf.*

Activity: Sand Movers

Objectives

- To explore how water can move sand
- To use evidence to back up children's scientific claims

Materials

- Sandbox, sensory table, or large watertight box
- Sand
- Water
- A variety of tools and containers to move water and create water flow such as watering cans, droppers, turkey basters, and empty condiment containers.
- Bucket for wastewater (if activity is indoors)
- Indirectly vented chemical-splash goggles
- Drawing or writing materials
- Camera (optional)
- Towels or mop for cleanup (optional)
- Smocks (optional)

Procedure

1. Draw children's attention to where rainwater or another source of water has moved sand or soil, if you can find such a location. This might occur on a dirt hillside or athletic field after a rain, along a stream bank, in a garden bed, or in a sandbox. You might say, *This curving pattern of soil looks like it was moved here somehow* or *I wonder why the sand is lower here?* to begin a discussion.

2. As children make guesses about how the sand or soil shape was formed, note any comments about prior experience. Plan to provide materials that will build on what the children have already experienced.

3. Provide a box of sand that can be used with water indoors, or use an outdoor sandbox. Match the size of the sandbox with the size of the containers the children will use to add water so water will not build up too quickly in the box. Wet sand can clog a sink, so if it is not possible to do this activity

outside, provide a bucket to carry waste water outside. *CAUTION: Have children wear indirectly vented chemical-splash goggles, and provide smocks if the children must stay dry.* Tell the children that you would like them to use just water to move the sand. (Encourage water-only movement of sand, but expect and accept it if children begin using their hands and the tools directly on the sand at some point.)

4. Allow children to explore the sand and water tools together for a long period of child-directed play, one hour or more a week depending on the interests of the children.

5. As children use the tools with water to move the sand and create shapes, ask open-ended questions to help them focus on how the force of moving water moves the sand such as *What happens to the sand when you do ____?*

6. Have children draw or photograph the shapes they make in the sand, and write or dictate a description of what tool(s) they used and how they made the shape.

7. Tell the children that the group is going to share what they know about the question, "Can water move sand?" Have children share their drawings and photos and talk in turns about their observations and experiences. If they disagree with a statement, say, *Scientists don't always agree with each other. We can learn from each other. What evidence do you have? Can you show what you drew or wrote?*

Children may want to demonstrate their actions. Keep the water and sandbox available for further exploration as long as the children are interested.

Teacher's Picks

Publications

Sand Dune Daisy: A Pocket Mouse Tale by Lili DeBarbieri, illustrated by M. Fred Barraza (Westcliffe, 2015).

Sand can be moved by wind as well as water. This book describes the desert dune environment where a pocket mouse lives among sands shifted by wind.

Is Sand a Rock? by Ellen Lawrence (Science Slam! Rock-ology; Bearport, 2014).

This book about the origin of sand describes how it is moved by water and wind.

Erosion by Becky Olien (Capstone Press, 2002).

Photographs illustrate and text describes the effects of erosion, at an early childhood level.

Website

National Geographic Education Encyclopedia: Erosion
http://education.nationalgeographic.com/education/encyclopedia/erosion/?ar_a=1

This entry is suitable for middle school students through adult learners.

Related *Early Years* Blog Posts

The following information is from "What Science Happens in Your Sandbox?" published April 19, 2013:

> *A pile of sand, a sandbox, and a sensory table full of sand are tools for imaginative play, sensory exploration, and science investigations. As children work with dry and wet sand, they notice and make use of the differences due to the properties of water:*
>
> * *Wet sand sticks together and can be made into deep holes and tall "mountains."*
>
> * *Footprints and other impressions are easy to make and see in wet sand.*
>
> * *Dry sand can slide off a shovel and flow into a hole or fill a bucket.*
>
> *The water molecules adhering to sand grains and each other aren't visible to the children, but they can explore this property and think about how that is the same or different from the way other materials behave.*

Read more at *http://nstacommunities.org/blog/2013/04/19/what-science-happens-in-your-sandbox.*

Additional related blog post:

"Soil Erosion in Miniature," published May 13, 2013: *http://nstacommunities.org/blog/2013/05/13/soil-erosion-in-miniature.*

Chapter 53

The STEM of Inquiry[54]

BEFORE YOU BEGIN THIS ACTIVITY:

Remember that each activity is not a stand-alone science or engineering curriculum. Activities are small steps in a journey of science inquiry, as discussed in Chapter 1. Your students will learn more about this concept and about the nature of science if you use this activity as part of an ongoing exploration of a question, a concept, or a topic being investigated by your class. Ask yourself, "What should come before this and what should come after?" Refer to Table 1.1 (pp. 15–39) to find other activities from The Early Years column that address the same concepts.

Many early childhood science lessons have children making observations of an organism or phenomena and documenting their observations (science). Counting and taking measurements may be involved (mathematics), and students may use tools to observe and document (technology). However, engineering concepts are often left out when they could easily be included. Planning to include an engineering process as part of an investigation can help us see how engineering can be an important part of the investigation—not a short activity tacked on to satisfy the "E" in STEM.

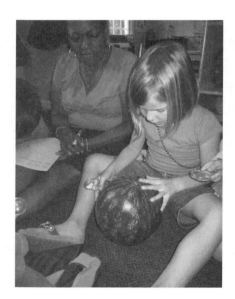

A science and engineering investigation may become an inquiry over time as children ask many related questions, investigate, collect data, and talk about their ideas. Deep exploration of concepts over time will allow students to use science and engineering practices and build toward mastery as their understanding matures. Children learn more in depth when they—with teacher support—are able to pursue areas of interest (Moomaw 2013). Just as it is likely that additional questions to investigate will come up as part of a science inquiry, the engineering design

---

[54] This column entry was originally published in *Science and Children* in October 2013.

process is iterative—the process can continue as students continue to ask, imagine, plan, create, improve, and so on.

The Engineering is Elementary program, developed by the Museum of Science, Boston (*www.eie.org*) has resources for elementary students that prekindergarten teachers can use to learn about the engineering design process. While reading about how older children investigate an engineering design problem (Bricker 2009), consider what is developmentally appropriate for your students. In the *Next Generation Science Standards*, K-2-ETS1 Engineering Design for kindergarten through grade 2 states that students who demonstrate understanding can "ask questions, make observations, and gather information about a situation people want to change to define a simple problem that can be solved through the development of a new or improved object or tool" and

"develop a simple sketch, drawing, or physical model to illustrate how the shape of an object helps it function as needed to solve a given problem" (NGSS Lead States 2013). Children's explorations and problem-solving before kindergarten—such as building with blocks and creating with tape, cardboard tubes, and paper—are foundations for later learning.

Plan for an activity that can inspire you and your students to work on a question or problem over a long period of time—enough time to allow questions to be fully explored to the satisfaction of young children. Children take on many tasks that involve solving a problem—for example, building a tower of small blocks when all the large blocks have been used by others. Some of their self-appointed problem-solving tasks may suggest a topic for future investigation by the entire class.

References

Bricker, M. 2009. Plants on the move. *Science and Children* 46 (6): 24–28.

Moomaw, S. 2013. *Teaching STEM in the early years: Activities for integrating science, technology, engineering, and mathematics.* St. Paul, MN: Redleaf Press.

NGSS Lead States. 2013. K-2-ETS1 Engineering Design. In *Next Generation Science Standards: For states, by states.* Washington, DC: National Academies Press. *www.nextgenscience.org/k-2ets1-engineering-design.*

Activity: Heavy Lifting

Objective

To investigate how to carry and transport a heavy object, including designing and using a tool

Materials

- Heavy objects (e.g., a pumpkin, large water bottle, bag of sand)—one object per small group

- Magnifiers

- Measuring tools (standard or nonstandard)

- Weighing scale

- Camera (optional)

- Various materials and tools familiar to the class that can be used as part of a system for transporting the heavy object (e.g., scarves, boards, baskets, cardboard boxes, scooter boards, toy vehicles, tape and scissors, short pieces of rope or yarn)

Procedure

1. As you read through this activity, note how science, technology, engineering, and mathematical concepts are all involved. Plan to include this activity as part of a larger inquiry or project such as one about simple machines, motion, or transportation.

2. Introduce the heavy object to your class by having students pass it around the circle to become familiar with all of its attributes. Challenge them to describe it so you can send the description home to their families. The name of the object will not be on the note; family members can have fun guessing what the object is. Remind the children that the name is going to be a secret. Children may not be able to describe more than color, shape, and other familiar attributes. Prompt them to also describe the texture, smell (if any), and weight and size (relative or measured) and to observe the motion when gently pushed.

3. As the children work, observe aloud that they are working as scientists when you see them using their senses and describing, and that they are using math

when they measure. Record, or have students record, the descriptions and measurements with text and drawings, according to their development.

4.   Introduce another challenge: to move the object safely (no breakage, bruising, spillage, or scrapes) from one place to another (locations chosen by you) without simply carrying it in their hands. Comment that engineers draw their ideas for solving problems like this one when they design structures and devices. Through drawing, children may consider additional design ideas—a habit of designing that can lead to success.

5.   Provide a variety of materials that students can use to create a carrier, in small groups or at a center. Assist students in cutting or tying materials, support their thinking about what is or is not working with comments to focus their attention (without telling them how to do it), and prompt them to redesign and rebuild if they appear frustrated. Ask students to explain their choice of materials to support their consideration of the properties of the materials and any alternatives.

6.   Children can show that their idea successfully moves the heavy object by demonstrating to the small group or by documenting with pictures, photographs, or video.

If a design does not successfully move the object, ask, *Were there any problems moving the object?* and *What one feature or part of your carrier could you change to make it work?* Noticing a problem with a design and changing one thing for each new test will help students understand that redesigning is expected and can determine which single change was needed to make the carrier work. Review the children's drawings or photographs and talk about how each carrier design transported the object safely. Write a note to send home using the students' descriptions.

Read a book that extends the learning about making observations; measuring, making, or improving a tool; or solving a problem (see "Teacher's Picks" section for additional suggestions for STEM learning).

Teacher's Picks

"Gimme an 'E'! Seven Strategies for Supporting the 'E' in Young Children's STEM Learning" by Cynthia Hoisington and Jeff Winokur, in *Science and Children* 54 (1): 3–10, 2015.

In this article, the authors share seven overlapping and mutually reinforcing strategies used by teachers to effectively support children's learning in physical science and engineering.

Family Engineering by Mia Jackson, David Heil, Joan Chadde, and Neil Hutzler (Foundation for Family Science and Engineering, 2011).

This is a book of fun, hands-on activities for elementary-age children and their families to explore engineering concepts and a guide for planning a family engineering event.

Have You Ever Seen a Stork Build a Log Cabin? by Etta Kaner, illustrated by Jeff Szue (Kids Can Press, 2010).

Illustrations of diverse animal-built and human-built structures highlight how other animals' structures can be a source of ideas for human construction.

Engineering the ABC's: How Engineers Shape Our World by Patty O'Brien Novak, illustrated by Don McLean (Ferne Press, 2009).

After reading this introduction to engineering, children will be asking, "How do you think an engineer made that?" about the everyday objects in their lives.

Eliza and the Dragonfly by Susie Caldwell Rinehart, illustrated by Anisa C. Hovemann (Dawn Publications, 2004).

This story includes an example of how a tool can solve a problem—how to see underwater.

How Tall, How Short, How Faraway by David Adler, illustrated by Nancy Tobin (Holiday House, 1999).

Nonstandard measuring units from historical Egypt and Rome led to customary measuring units, as explained in this book that introduces the concept of measuring.

Bridges Are to Cross by Philemon Sturges, illustrated by Giles Laroche (Putnam Juvenile, 1998).

The many examples of bridges in this book inspire young builders to create their own. Bridges solve the problem stated in the title, "to cross."

Galimoto by Karen Lynn Williams, illustrated by Catherine Stock (HarperCollins, 1991).

Using found materials, a child creates a toy with moving parts.

Related *Early Years* Blog Posts

The following information is from "Found Materials + Engineering Process = Toy," published April 26, 2012:

> *We didn't have enough wire so we reused cardboard tubes, empty boxes, egg cartons, and plastic jar lids to create toys called galimotos in the Malawian children's tradition as recounted in the children's book* Galimoto *by Karen Lynn Williams and illustrated by Catherine Stock (HarperCollins, 1991).* Galimoto *means "car" in Chichewa, the national language of Malawi, and many of our creations were vehicles. The small group of kindergarten and first-grade girls drew their designs and then built their own toy to take home in a one-hour library-sponsored program.*

Read more at *http://nstacommunities.org/blog/2012/04/26/found-materials-engineering-process-toy*.

Additional related blog post:

"DiscoverE Brings Us Engineers Week," published February 11, 2014: *http://nstacommunities.org/blog/2014/02/11/discovere-brings-us-engineers-week*.

Chapter 54

Are They Getting It?[55]

BEFORE YOU BEGIN THIS ACTIVITY:

Remember that each activity is not a stand-alone science or engineering curriculum. Activities are small steps in a journey of science inquiry, as discussed in Chapter 1. Your students will learn more about this concept and about the nature of science if you use this activity as part of an ongoing exploration of a question, a concept, or a topic being investigated by your class. Ask yourself, "What should come before this and what should come after?" Refer to Table 1.1 (pp. 15–39) to find other activities from The Early Years column that address the same concepts.

All teachers want to know if their students are learning the concepts that we hope we are teaching them. In early childhood, development varies widely between children of the same age, and decisions about teaching should be as individualized as possible (NAEYC 2009). To individualize curriculum and teaching, teachers must first know what their students already think and know. By knowing what ideas the children already have about a concept, teachers can plan learning experiences and discussions to help build further understanding.

Formative assessments that check on student learning as an inquiry or project progresses help us determine what children understand so we can use additional strategies to address unanswered questions. Formative assessment can be as simple as writing down children's words as they talk with other children at the sensory table or noting which children use the vocabulary word *wet* correctly. Chalufour and Worth (2006) call on teachers to watch children's interactions with materials and with each other and use their observations as "the basis for selecting interventions that are relevant to what the children are doing and thinking and that improve their understanding" (p. 104).

Working with materials is the focus of the following activity, relating to disciplinary core idea PS1.A, which states that by the end of grade 2, students should know that matter can be described by its observable properties and that different properties are suited to different purposes (NRC 2012). Experiences with a wide variety of materials for pouring, transferring, weighing, bouncing, stacking, balancing,

---

[55] This column entry was originally published in *Science and Children* in November 2013.

cutting, taping, gluing, stapling, and tying teach children about the properties of "wetness," texture, weight, stretch, strength, absorption, flexibility, and adherence. All that knowledge can develop from having access to materials for "free art" *and* an observant teacher to ask open-ended questions focusing on, and extending, what the children notice about a material.

In the following activity, children gain experience in gauging the amount or volume of objects and observing the properties of materials such as wax paper and cardboard. They also become engaged in designing a container that will hold and transport their favorite food. Prior knowledge of carryout pizza boxes and Styrofoam cold drink cups may lead them to ask, "What is it about those containers that make them useful for transporting those foods?"

Engineering practices (NGSS Lead States 2013) are used to define engineering problems (such as creating a container), design solutions, and optimize the design solution. The exploration of materials and the investigation into making containers can be open-ended, at a station or center, and during discussion time in individual conversations or at a morning meeting or closing circle. Throughout, teachers use formative assessment to find gaps in understanding and plan additional instruction.

References

Chalufour, I., and K. Worth. 2006. Science in kindergarten. In *K today: Teaching and learning in the kindergarten year,* ed. D. F. Gullo, 95–106. Washington, DC: National Association for the Education of Young Children (NAEYC). Also available online as Reading #56. Science in kindergarten. A reading from the CD accompanying *Developmentally appropriate practice in early childhood programs: Serving children from birth through age 8,* 3rd ed., eds. C. Copple and S. Bredekamp. Washington, DC: NAEYC, 2009. *www.rbaeyc.org/resources/Science_Article.pdf.*

National Association for the Education of Young Children (NAEYC). 2009. *Position statement: Developmentally appropriate practice in early childhood programs serving children from birth through age 8.* Washington, DC: NAEYC. *www.naeyc.org/files/naeyc/file/positions/PSDAP.pdf.*

National Research Council (NRC). 2012. *A framework for K–12 science education: Practices, crosscutting concepts, and core ideas.* Washington, DC: National Academies Press. *www.nap.edu/catalog/13165/a-framework-for-k-12-science-education-practices-crosscutting-concepts.*

NGSS Lead States. 2013. Appendix I—Engineering design in the *NGSS.* In *Next Generation Science Standards: For states, by states.* Washington, DC: National Academies Press. *www.nextgenscience.org/next-generation-science-standards.*

Activity: Materials Have Properties

Objectives

- To experience the properties of a wide variety of materials

- To use understanding of the properties of materials to make a container that successfully contains a food item

Materials

- Sensory table or tub with various containers for exploring the concept of volume

- A variety of measuring tools

- A wide variety of materials: fabric; a variety of papers (newspaper, paper towels, construction paper, and wax paper); plastic wrap; cardboard sheets and tubes; pieces of wood; aluminum foil; string

- Scissors

- Stapler, tape, and glue

Procedure

1. To provide foundational experiences, plan for children to have long periods of time at a sensory table or tub to explore the concept of volume using various sizes and types of containers such as cups, scoops, plates, sieves, colanders, boxes, and tubes. Put water and, later, sand or rice in the sensory table to provide experience with both dry and wet materials. Observe students to assess if they are selecting appropriate size and shape containers for an amount of water or sand and using appropriate and detailed vocabulary to describe materials. If not, provide some of the activities and discussion to improve understanding. Also provide extended open-ended exploration time with a collection of materials to "make something," sometimes called free art or "invention exploration."

2. Talk about the properties of food during a mealtime. For example, you could make one or more of these statements: *This food is in small pieces. My drink is wet; it's a liquid. My fingers are greasy from holding the pizza. I had to cut the celery into pieces to fit into my lunch box.*

3. Have children name their favorite drink or food eaten at school, describe it, and explain what they like about it. Record descriptions and assess by noticing if the children use accurate terms for the food selected—pizza: *greasy, flat, solid*; juice: *wet, liquid*; ice cream: *cold, solid-then-liquid*. If not, provide additional activities to experience different shapes and states of matter and discuss differences and learn vocabulary words.

4. Ask children to describe containers we use for eating food that work well to hold the food, keep us clean, keep heat or cold in, and allow food to be transported without spilling. Children might describe how bowls keep cereal and milk together or how a carton of milk keeps the milk from spilling. They can describe the size of the container in relative terms or measure with a tool. Teachers can provide a list of descriptive words for food, such as *warm, cold, wet, dry, greasy, breakable, ice*, and for the container: *big, small, tall, cup, lid, handle, flat, round, paper, plastic, wood*, and *metal*.

5. Have children draw a picture of a "good" container to hold their favorite food—making changes to a familiar container or inventing one. The size of the "good" container can be described in relative terms or measured with a tool. Help children describe and label the container's parts and say what they do or why they are needed (e.g., a handle to carry or a seal to hold the bag closed). Do the children consider whether the container will hold a liquid or a solid, will hold a hot or cold food and drink, and will have a shape tall, wide, or big enough to hold their food choice?

6. Revisit the materials collection and have students talk about which materials will be best for making their container, and then have them build it. Do they choose materials with appropriate properties to build a container able to hold their choice of food? If not, provide additional activities to test materials for ability to insulate, resist leakage, and maintain strength. Teachers can be the "hands" of children who need help cutting and taping, but you should not direct the construction.

When children use their container, does the design successfully hold the food item? If not, encourage students to analyze the problem and consider changes to their design. Some students may need support through a discussion of how we can learn from designs that don't work and encouragement to consider alternatives.

Teacher's Picks

"Science 101: Q: What Is the Difference Between Solids and Liquids?" by Bill Robertson, in *Science and Children* 50 (9): 72–73, 2013.

Robertson discusses the distinction between solids and liquids and describes some examples where it is not clear-cut.

Amazing Materials by Sally Hewitt (Amazing Science; Crabtree, 2008).

The photographs of common materials in this book help explain how materials can be solid, liquid, or gas. Understanding properties of materials will help children design containers that work as they intended.

Change It! Solids, Liquids, Gases and You by Adrienne Mason, illustrated by Claudia Dávila (Primary Physical Science; Kids Can Press, 2006).

This book helps children understand the properties of states of matter by describing how shapes can be changed through pushing, pulling, breaking, pouring, melting, freezing, or evaporation. It includes hands-on activities to do at school or at home.

I Get Wet (Vicki Cobb Science Play) by Vicki Cobb, illustrated by Julia Gorton (HarperCollins, 2002).

The concept of adhesion (sticking to itself and other materials) is introduced with activities for children to try.

What Is the World Made Of? All About Solids, Liquids, and Gases by Kathleen Weidner Zoehfeld, illustrated by Paul Meisel (Let's-Read-and-Find-Out Science, Stage 2; HarperCollins, 1998).

Illustrations of children engaging in exploration and discussion about matter help connect the discussion of science concepts with everyday life experiences.

The following three works of fiction are good to read aloud when children raise questions about why people use containers for serving food and the kinds of containers and tools for eating food used in other cultures:

How Do Dinosaurs Eat Their Food? by Jane Yolen, illustrated by Mark Teague (Blue Sky Press, 2005).

After reading this book, children can talk about, draw, or design a food container that is appropriate for an animal as large as a dinosaur.

First Book of Sushi by Amy Wilson Sanger (World Snacks; Tricycle Books, 2001) and other books in the World Snacks series.

This series shows how many different kinds of foods are served and contained.

Cloudy With a Chance of Meatballs by Judi Barrett, illustrated by Ron Barrett (Atheneum, 1978).

This humorous story involves people dealing with solid and liquid foods falling from the sky.

Related *Early Years* Blog Posts

The following information is from "Problem Solving and Investigating the Properties of Materials," published November 13, 2013:

> *Do the approaches and strategies a child uses for finger painting or eating a messy snack tell us anything about how she or he will approach building with blocks or participating in a science activity? There are problem-solving tasks in all of these activities. If we tell children how to do a task, they may not discover other ways, or the best way for their style.*

Read more at *http://nstacommunities.org/blog/2013/11/13/problem-solving-and-investigating-the-properties-of-materials.*

Additional related blog posts:

"First Day of School Science," published August 2, 2011: *http://nstacommunities.org/blog/2011/08/02/first-day-of-school-science.*

"Concepts That Cut Across Science Disciplines," published January 26, 2014: *http://nstacommunities.org/blog/2014/01/26/concepts-that-cut-across-science-disciplines.*

Chapter 55

Now We're Cooking[56]

BEFORE YOU BEGIN THIS ACTIVITY:

Remember that each activity is not a stand-alone science or engineering curriculum. Activities are small steps in a journey of science inquiry, as discussed in Chapter 1. Your students will learn more about this concept and about the nature of science if you use this activity as part of an ongoing exploration of a question, a concept, or a topic being investigated by your class. Ask yourself, "What should come before this and what should come after?" Refer to Table 1.1 (pp. 15–39) to find other activities from The Early Years column that address the same concepts.

At daily meals, children have ample opportunity to learn about the categories of whole and half, cooked and raw, and solids and liquids. They can observe how some foods seem to be neither solid nor liquid or can change state from one to another ("My ice cream melted!"). Young children are often very interested in cooking. It is an activity that is familiar to them from home, but they may not have participated in cooking yet. They flock to the table to cut apple slices into smaller pieces to cook applesauce or to mix dry and wet ingredients together to make dough, and they like to use a freezer to change the temperature of other mixtures.

Cooking may involve observing, counting, using parts of plants, examining the properties of materials, selecting the appropriate tool for a job, and making changes. These activities provide experience with disciplinary core ideas described in *A Framework for K–12 Education* (*Framework*; NRC 2012), including PS1.A ("Different kinds of matter exist and many of them can be either solid or liquid, depending on temperature. Matter can be described and classified by its observable properties.") and PS1.B ("Heating or cooling a substance may cause changes that can be observed."). Making changes to materials as part of cooking can build understanding toward

---

[56] This column entry was originally published in *Science and Children* in December 2013.

the grade 2 performance expectation 2-PS1-4: "Construct an argument with evidence that some changes caused by heating or cooling can be reversed and some cannot" (NGSS Lead States 2013). Children's experiences and observations will be their evidence for which changes can, or cannot, be reversed. As seen in the following activity, cooking also uses many of the practices and skills used in science and engineering investigations.

If you find that some children never come to the cooking table, observe them closely to see if you can figure out what might be a barrier for their participation. Some children do not like to get messy—if this is the issue, offer them the job of assembling the tools, or give them smocks to protect their clothes. Some children want to "make their own"; to address this preference, give each child a small cup to mix ingredients together before putting all the batter together in the baking dish. Some children are unsure of what will be required of them in the process; to help them, prepare a recipe card with pictures showing the actions of each step. If certain children are not drawn to the cooking activities you plan, consider making a recipe that relates to a favorite book (see "Teacher's Picks" section), or invite a parent to come in and help the children prepare a favorite family dish. Simple exploration of vegetables and cutting them into smaller pieces can be a first step into the science concepts in cooking (Kalich, Bauer, and McPartlin 2009).

CAUTION: Young children especially like to explore the world with their mouths, so early childhood science explorations are a good time to begin teaching the safety rule of "No tasting in science." Call your science activities and investigations with food "cooking," and explain that all the ingredients you are giving children to explore are safe to eat.

References

Kalich, K. A., D. Bauer, and D. McPartlin. 2009. Early sprouts: Establishing healthy food choices for young children. *Young Children* 64 (4): 49–55.

National Research Council (NRC). 2012. *A framework for K–12 science education: Practices, crosscutting concepts, and core ideas.* Washington, DC: National Academies Press. *www. nap.edu/catalog/13165/a-framework-for-k-12-science-education-practices-crosscutting-concepts.*

NGSS Lead States. 2013. *Next Generation Science Standards: For states, by states.* Washington, DC: National Academies Press. *http://nextgenscience.org/2ps1-matter-interactions*.

Activity: Ice Cream Science

Objective

To make observations of the ingredients for ice cream before mixing them together and after making a change by cooling the ingredients

Materials

For a class of 18 to each have a small cup of ice cream:

- Two pint-size and two gallon-size sealable (zip close) plastic freezer bags, or other sealable containers that fit one inside another with room for ice on all sides

- Wide-width tape

- Two pairs of children's mittens

- 1 pint cream

- 1 cup milk

- 6 tablespoons sugar

- 1 teaspoon vanilla extract

- Fresh or frozen fruit (optional)

- 8 cups ice (about three or four ice-cube trays)

- $^2/_3$ cup salt (table or kosher)

- Small serving cups and spoons

- Smocks (optional)

- Thermometer safe for children to use (optional)

Teacher Preparation

There are many online sites with recipes for making ice cream using plastic bags or other containers such as coffee cans to hold the ingredients. Ahead of time, copy one of them or the recipe below onto a poster for easy viewing, drawing pictures or using photographs to represent the tools, ingredients, and actions. *CAUTION: Check student forms for allergies before bringing any food into the classroom.*

Procedure

1. In small groups or as a class, follow a recipe for making ice cream (see Box 55.1, p. 332, for one recipe, but you can use one from another source; the recipe in the box is for two containers of ingredients, to allow more students to participate at one time). Read the recipe aloud and say that as part of a cooking activity, the students will also be using the science practices of making and communicating observations, asking questions, and making measurements.

2. Help children measure the ice cream ingredients to go in the pint bag and the ice and salt for the gallon bag. Include time for touching and describing the ingredients.

3. Have the students feel the bag containing the ice cream ingredients, and then the bag containing the ice and salt, and describe their relative temperatures. Have them measure the temperature of the ingredients in the bag and the ice, if they are familiar with the use of a thermometer.

4. Have children wear mittens and gently but firmly massage and shake the bags. To keep busy for the 5–15 minutes needed for the ingredients to lose heat and become cold enough to form tiny ice crystals, pass the combined bags around a circle while the class sings songs, take them with you for a walk, or engage the children in preparing the serving cups and optional fruit, shaking the bags continuously.

5. Discuss the change in temperature and texture with the children, and measure with a thermometer if this is developmentally appropriate. Eat the ice cream and talk with the children about any changes due to melting as they snack.

After preparing and eating the ice cream, review the steps that were followed to make it, and the changes that occurred. By adding salt to the ice, the freezing temperature of the ice was lowered, allowing the ice cream ingredients to cool and freeze as well. This process is interesting to sense through touch with hands and tongue, and for older students to measure, preparing for deeper understanding of energy transfer in freezing and melting in later years.

> ### BOX 55.1. ICE CREAM RECIPE (makes 18 small servings)
>
> *Involve children in every step of the recipe.*
> 1. In each of the two pint-size sealable bags, add 1 cup of cream, ½ cup of milk, 3 tablespoons of sugar, and ½ teaspoon of vanilla extract. Completely dry the opening of the bag, push most of the air out and seal, then tape along the opening as a precaution.
> 2. In each of the two gallon-size bags, add 4 cups of ice and ⅓ cup of table or kosher salt.
> 3. Put each pint bag of ice cream ingredients into each gallon bag of ice and salt. Dry and seal the opening of the gallon bag, and tape along the opening.
> 4. Gently but firmly massage and shake the bags.
> 5. Feel the inner bag to assess the texture of the ingredients. When the mixture feels more solid, the ice cream is ready to eat. Cut open the gallon bags, rinse the salty water off the pint bags, and cut open one corner to squeeze a serving into each cup.

Teacher's Picks

Publications

The first seven children's books listed below include recipes:

Rainbow Stew by Cathryn Falwell (Lee & Low Books, 2013).

Children search in the garden for food in the colors of the rainbow to make a stew with Grandpa. Your class could make such a colorful stew!

Tea Cakes for Tosh by Kelly Starling Lyons, illustrated by Earl B. Lewis (G.P. Putnam's Sons, 2012).

This book shows a loving connection across generations. Eating Grandma Honey's tea cakes and listening to her family history links a child to his ancestors' struggles, survival strategies, and resilience—strengths the child uses himself as Honey ages and becomes forgetful.

Soup Day (Christy Ottaviano Books) by Melissa Iwai (Henry Holt, 2010).

Make cooking soup a weekly event in your program; try this book's recipe for Snowy Day Vegetable Soup.

Honey Cookies by Meredith Hooper, illustrated by Alison Bartlett (Frances Lincoln Children's Books, 2005).

This book provides an easy recipe for making cookies from scratch, as well as an education in where butter, sugar, flour, eggs, cinnamon, and honey come from.

The Story of Noodles by Ying Chang Compestine, illustrated by YongSheng Xuan (Holiday House, 2002).

This origin story of how a recipe gone wrong became a new way to shape a staple food may inspire your students to try new ways to shape materials such as playdough.

Cook-a-doodle-doo! by Janet Stevens and Susan Stevens Crummel (Harcourt Brace, 1999).

Descriptions of kitchen tools and the source of ingredients fill sidebars in this book with information to support the questions children have after hearing the very funny story.

The Ugly Vegetables by Grace Lin (Charlesbridge, 1999).

In this book, the benefit of growing vegetables, even "ugly" ones like bitter melon, becomes clear to the child when she smells the soup aroma. In a garden, children can search for the flowers that precede the development of the vegetables.

The next three books highlight science concepts in the kitchen, with recipes for edible and non-edible explorations:

Kitchen Science by Shar Levine and Leslie Johnstone (Sterling, 2003).

No recipe is included in the book, but raw ingredients for both vegetable and fruit salads are described in the text.

Oliver's Vegetables and *Oliver's Fruit Salad* by Vivian French, illustrated by Alison Bartlett (Orchard Books, 1995 and 1998).

Investigating a vegetable garden leads Oliver to investigate the taste of vegetables. Making a fruit salad makes tasting fruits a delight.

The last four books describe how ice cream is made:

Ice Cream: The Full Scoop by Gail Gibbons (Holiday House, 2006).

Milk to Ice Cream by Inez Snyder (Welcome Books: How Things Are Made; Children's Press, 2003).

Ice Cream by Elisha Cooper (HarperCollins, 2002).

From Cow to Ice Cream by Bertram T. Knight (Children's Press, 1997).

Website

Vanilla Ice Cream Without the Machine
www.seriouseats.com/recipes/2010/07/vanilla-ice-cream-without-the-machine-recipe.html

For adults who want to try more involved ice cream recipes at home, the Serious Eats website offers this recipe by chef and writer J. Kenji López-Alt.

Related *Early Years* Blog Posts

The following information is from "Cooking as Science," published December 31, 2013:

> *In her "What About Food and Cooking?" post on the Pondering Preschool blog (published May 14, 2013; http://myiearlychildhoodreflections.blogspot. com/2013/05/what-about-food-and-cooking.html) Maureen Ingram tells how her class of 3-year-olds used early literacy skills and design skills as they developed imaginative play extensions of a nutrition and cooking project. They also used engineering practices as they designed and built devices to get a great big turnip out of the ground, inspired by a traditional Russian tale.*

Read more at *http://nstacommunities.org/blog/2013/12/31/cooking-as-science.*

Additional related blog post:

"I Had a Carrot for Breakfast," published July 17, 2009: *http://nstacommunities.org/ blog/2009/07/17/i-had-a-carrot-for-breakfast.*

Chapter 56

Inviting Parents to School[57]

BEFORE YOU BEGIN THIS ACTIVITY:

Remember that each activity is not a stand-alone science or engineering curriculum. Activities are small steps in a journey of science inquiry, as discussed in Chapter 1. Your students will learn more about this concept and about the nature of science if you use this activity as part of an ongoing exploration of a question, a concept, or a topic being investigated by your class. Ask yourself, "What should come before this and what should come after?" Refer to Table 1.1 (pp. 15–39) to find other activities from The Early Years column that address the same concepts.

Parents and other adults in the families of young children can be a rich resource for the science and engineering programs in preschools, home day care programs, child care centers, and elementary schools. Adults like to be classroom heroes for their children. Holding a brief discussion with children or reading a book aloud at the beginning or end of the school day may fit into the schedule of parents with inflexible working hours. Adults who have more flexibility can be scheduled visitors with specific agendas to support the current science or engineering investigation:

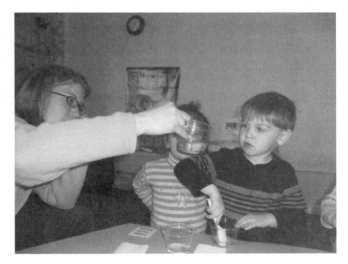

- A plumber can bring in (cleaned) tools and allow the children to open and close an adjustable pipe wrench as part of an exploration into building.

- A cashier can participate in imaginary play and help children classify and sort play money.

[57] This column entry was originally published in *Science and Children* in July 2014.

- A musical parent can bring in a guitar to play for a class sing-along as part of an exploration of sound.

- An engineer can share a real-life problem with the class and ask the children to draw some possible solutions.

- A scientist can bring in specimens of fossils she is investigating for the children to touch and wonder about.

Whether parents are a regular part of your early childhood program or occasional guests, rather than have them simply observe, you can make them feel wanted and needed beyond their specialty by involving them in the work the children do. Adults who are not teachers often need instruction on asking the children open-ended questions rather than telling them what is happening or what the adult knows. As beginning teachers, none of us were masters at listening to children explain their ideas before sharing what we know. Parents and other adults will also need practice (and your gentle direction) to become guides who "allow children to question, explore, investigate, make meaning, and construct explanations and organize knowledge by manipulating materials" (NSTA 2014).

The following "cooking" activity—making lemonade from scratch—is an experience where children can touch a liquid (water), feel a granular solid (sugar), smell and taste a lemon, measure out ingredients, and observe the process of sugar dissolving in water. Children often confuse dissolving with melting because in both processes matter seems to "disappear." Investigating the observable properties of materials is part of the second-grade performance expectation 2-PS1-1: "Plan and conduct an investigation to describe and classify different kinds of materials by their observable properties" (NGSS Lead States 2013). Visiting adults can support children's experience and discussion of the science concepts involved.

References

National Science Teachers Association (NSTA). 2014. NSTA position statement: Early childhood science education. *www.nsta.org/about/positions/earlychildhood.aspx*.

NGSS Lead States. 2013. *Next Generation Science Standards: For states, by states.* Washington, DC: National Academies Press. *http://nextgenscience.org*.

Activity: Making Lemonade From Scratch

Objective

To engage children in observing the process of a solid dissolving into a liquid as they mix the two together

Materials

- Safety goggles for each participant

- Pitchers small enough for the children to handle successfully

- Water at room temperature

- Sugar, 1 teaspoon per child

- Measuring cups and spoons

- Small clear cups, one per child

- Spoons or craft sticks, one per child

- Lemons (washed), enough for a small slice per child

- Napkins (optional)

Teacher Preparation

Check with families beforehand for any food allergies or dietary restrictions.

Procedure

1. Review the procedure with any adult assistants. Remember to mention safety considerations such as washing hands before starting the activity and wearing safety goggles during the activity. Give visitors a list of suggested open-ended questions to promote student observation and thinking during the activity.

2. With adult assistance, have children pass out and put on safety goggles, help to assemble ingredients, fill small pitchers with room-temperature water, and put cups and spoons for each child on the table. *CAUTION: Remind students not to share spoons and cups.* Because this activity culminates in tasting the mixture (lemonade), tell children that it is a cooking activity that involves science concepts, rather than a science activity.

3. Pour water into your spoon and taste it before pouring water in the children's spoons for them to taste. Children can discuss questions like, "How does the water feel and taste?" in smaller groups with an adult helper.

4. Repeat the tasting process with small amounts of sugar on the adults' and children's spoons. Have children describe what they see, feel, and taste.

5. Have adults model measuring and pouring about ½ cup of water and a teaspoon of sugar into the cup and observing it without stirring. Then help children do the same in their cups. Tell the children they may taste it later when all the mixing is finished. Ask, *How much sugar is on the bottom of your cup?* Children may say, "All of it!" or "It's all over the bottom."

6. To help children notice changes after a minute or two of stirring, say, *Tell me how much sugar you can see in your cup now.* Children may say, "The sugar disappeared!" Children can learn that the sugar is still there by tasting a spoonful of the solution. Model tasting your solution and say, *Discuss with your group what the liquid in your cup tastes like now.*

7. Provide lemon slices for children to add to their cups to make lemonade. While drinking, ask the children to recall the process they followed and the ingredients they used to make the lemonade. With the help of an adult, they can even write and illustrate a book about the process.

Pair this activity with others where children can make mixtures. NSTA's position statement on early childhood science education (see Appendix 2) emphasizes, "To effectively build science understanding, young children need opportunities for sustained engagement with materials and conversations that focus on the same set of ideas over weeks, months, and years." Repeated mixing of various safe and common solids and liquids, such as sand and water or salt and water, will help children build understanding that some solids dissolve and others do not and identify this process as being different from melting.

Teacher's Pick

Mixing and Separating by Chris Oxlade (Changing Materials; Heinemann, 2009).

This book defines "mixture" and provides examples of mixtures made by dissolving matter. Read this book to children to help them reflect on their own understanding, *after* they have hands-on experiences.

Related *Early Years* Blog Posts

The following information is from "Lemonade Stands and Summer Activity Resources," published July 9, 2014:

> *Now that the weather is hot, I see evidence of family support for science and math learning and review on neighborhood street corners, in the form of lemonade stands. "Made from scratch" lemonade requires cutting open fruit, squeezing the juice out of all the little vesicles in the lemons, measuring water and sugar, and stirring to mix it all together. With a magnifier, children can make close observations of the structure of a lemon, looking closely at the seeds and juice-holding vesicles and at grains of sugar. Even very young children can learn the vocabulary words* wet *and* dry, liquid *and* solid, *and* dissolve.

Read more at *http://nstacommunities.org/blog/2014/07/09/lemonade-stands-and-summer-activity-resources*.

Additional related blog post:

"Summer Reading, Summer Camping, Summer Science," published June 9, 2010: *http://nstacommunities.org/blog/2010/06/09/summer-reading-summer-camping-summer-science*.

Chapter 57

The Nature of Science in Early Childhood[58]

BEFORE YOU BEGIN THIS ACTIVITY:

Remember that each activity is not a stand-alone science or engineering curriculum. Activities are small steps in a journey of science inquiry, as discussed in Chapter 1. Your students will learn more about this concept and about the nature of science if you use this activity as part of an ongoing exploration of a question, a concept, or a topic being investigated by your class. Ask yourself, "What should come before this and what should come after?" Refer to Table 1.1 (pp. 15–39) to find other activities from The Early Years column that address the same concepts.

"It's science! You're going to learn some cool facts!" I often hear this from parents who are commenting with enthusiasm about part of the pre-school curriculum. However, this outlook does not emphasize the nature of science or how science works. I would like parents to say, "Go see what you can find out!" to encourage investigative behavior in their children, rather than suggest that the teacher will tell them interesting facts.

Teachers of young children ask, "Should we only teach facts and offer experiences, or should we teach how science is conducted? Are children able to understand?" The Understanding Science website developed by the University of California Museum of Paleontology (*http://undsci.berkeley.edu/index.php*) states, "Kindergarten, first, and second grade students can begin to understand what science is, who does science, and how scientists work through classroom activities, stories about scientists, and class discussions" (Janulaw n.d.). Although "investigate questions and gather evidence through observations" and holding "class discussions to share evidence and ideas" are listed as appropriate for first grade, beginning efforts in these tasks can start in preschool with developmentally appropriate expectations. Children who have an understanding of the nature of science may be able to apply it to many situations, not only in familiar situations.

---

[58] This column entry was originally published in *Science and Children* in September 2014.

The nature of science is usually described as having six to eight aspects, including understanding the difference between observation and inference (Lederman and Lederman 2004; Quigley, Buck, and Akerson 2011). Read more about the nature of science in Appendix H of the *Next Generation Science Standards* (NGSS Lead States 2013). The following activity provides practice making observations and inferences. Paired with children's literature (see "Teacher's Picks" section) that directs chil-

dren to make and discuss inferences based on the clues, and many additional opportunities to make observations, children will begin to tell you what they observed and what they think it means. Supporting their ideas with evidence, including their inferences, children are using one of the science practices described in *A Framework for K–12 Science Education* (NRC 2012)—Engaging in Argument From Evidence.

References

Janulaw, S. n.d. Know your students: Implications for understanding the nature of science. University of California Museum of Paleontology. *http://undsci.berkeley.edu/teaching/k2_implications2.php*.

Lederman, N. G., and J. S. Lederman. 2004. Revising instruction to teach nature of science. *The Science Teacher* 71 (9): 36–39.

National Research Council (NRC). 2012. *A framework for K–12 science education: Practices, crosscutting concepts, and core ideas.* Washington, DC: National Academies Press. *www.nap.edu/catalog/13165/a-framework-for-k-12-science-education-practices-crosscutting-concepts*.

NGSS Lead States. 2013. Appendix H—Understanding the scientific enterprise: The nature of science in the *Next Generation Science Standards. Next Generation Science Standards: For states, by states.* Washington, DC: National Academies Press. *www.nextgenscience.org/next-generation-science-standards*.

Quigley, C., G. Buck, and V. Akerson. 2011. The nature of science challenge. *Science and Children* 49 (2): 57–61.

Activity: Observations and Inferences

Objective
To introduce the concept of creating inferences based on observations

Materials

- Two copies of a photograph or illustration of a common object or scene (8 ½ × 11 inch or larger for visibility)

- Pieces cut from one copy of the photo—*not* like a puzzle, but simply some pieces cut out

- The actual item pictured (optional)

Teacher Preparation

Print two copies of a photograph or illustration that shows a locally common object or scene. Cut out some pieces so that the portion shown is not easily identifiable. For example, from a photo of a dandelion, you could cut out sections of leaves, flower buds, and stems.

Procedure

1. With a small group of three to six children, show one piece from the larger picture, keeping the intact picture hidden. Say aloud, *I wonder what is shown in this picture?* The children may offer ideas about the object, but tell them that you are describing only exactly what you see, such as, *I see a green shape with lines going through it.*

2. Say that you are now going to make an inference—to say what you think the picture might be. State an inference that will not give away the identity of the object and is slightly off track, to allow you to model changing your understanding in light of new information later. For the example of a piece of a photo of a dandelion, you might say, *It looks like a tree because it is green and I have seen a lot of green trees.*

3. Distribute one piece of the picture to each child and ask the students to think about *what they see* in the picture. Use strategies such as "turn and talk" pairing or taking turns to have all children describe their pieces of the picture.

4. After all children describe their pieces of the picture, ask them to make an inference—to say what they think it might be—and why they think that. Tell children that sometimes scientists don't know what something is, and that is okay. For example, children may say, "leaves," "a tree," or "something you blow," depending on which part of the plant is in their pieces of the picture. The objective is not to guess what is shown in the whole picture but to teach children how to make observations that report

exactly what they see, and then to make inferences about what the object might be based on previous knowledge and observations.

5. As children state their inferences, have them put their pieces in the middle of the work space. Encourage children to think about why they made a particular inference by asking, *What made you think of that?*

6. Show the intact picture and tell children that their pieces are not parts of a puzzle but just a few sections from the whole, so they won't try to make their piece fit in with other pieces instead of matching it to the whole. Show where your piece matches the intact picture. Comment on how your inference, based on what you could see, did not correctly identify the object pictured: *Oh, it's not a tree!* Ask children to find spots on the large picture for the pieces they hold to help them see what parts are shown in the pieces. If available, provide the actual object pictured and have children make more observations using other senses.

Teacher's Picks

In the Garden: Who's Been Here? (2008); *Around the Pond: Who's Been Here?* (1996); *In the Snow: Who's Been Here?* (1995); and *In the Woods: Who's Been Here?* (1995) by Lindsay Barrett George (Greenwillow Books).

In each of the *Who's Been Here* books, readers follow two children on a walk in nature and can make inferences about the animals that live there based on the pictured evidence and their prior knowledge. Each page that follows the evidence shows "who."

Seven Blind Mice by Ed Young (Reading Railroad; Puffin, 2002).

In this retelling of an ancient Indian story, the mice infer what an elephant is, based on what they feel as they each touch a different part of the elephant. Pair this book with the activity about observations and inferences and ask your students to identify the inferences made by the mice.

Read the following three articles to refresh your understanding of inference and observation and get tips for teaching about the difference between these science process skills:

"Inference or Observation?" by Kevin D. Finson, in *Science and Children* 48 (2): 44–47, October 2010.

"Perspectives: Learning to Observe and Infer" by Deborah L. Hanuscin and Meredith A. Park Rogers, in *Science and Children* 45 (6): 56–57, February 2008.

"Science Shorts: Observation Versus Inference" by Craig R. Leager, in *Science and Children* 45 (6): 48–50, February 2008.

Related *Early Years* Blog Posts:

The following information is from "When Are Children Old Enough to Smell a Flower, Touch an Earthworm, or Talk About the Nature of Science (NOS)?" published September 4, 2014:

> *When are children old enough to begin exploring the natural world? Can a 3-year-old touch a crawling beetle? Can a 2-year-old smell a flower; can a 1-year-old? Can a 3-month-old feel a leaf? This question was raised in a recent training session about helping young children learn more about the small animals they are curious about: worms, insects, and other small animals. People disagreed about what age would be appropriate to allow children these experiences, but we agreed that 5-year-olds are usually ready to observe for longer periods and that early childhood educators must know the children in their program and what activities they can safely engage in.*

Read more about the nature of science at *http://nstacommunities.org/blog/2014/09/04/when-are-children-old-enough-to-smell-a-flower-touch-an-earthworm-or-talk-about-the-nature-of-science-nos.*

Additional related blog post:

"Early Childhood Science in Preschool—a Conversation on Lab Out Loud," published February 24, 2014: *http://nstacommunities.org/blog/2014/02/24/early-childhood-science-in-preschool-a-conversation-on-lab-out-loud.*

Appendix 1

Additional Resources for Science Inquiry and Engineering Design

Publications

Ansberry, K., and E. Morgan. 2007. *More picture perfect science lessons: Using children's books to guide inquiry, K–4*. Arlington, VA: NSTA Press.

Committee on K–12 Engineering Education; L. Katehi, G. Pearson, and M. Feder, eds. 2009. *Engineering in K–12 education: Understanding the status and improving the prospects*. Washington, DC: National Academies Press.

Finkelstein, A. 2001. *Science is golden: A problem-solving approach to doing science with children*. East Lansing: Michigan State University Press.

Gelman, R., K. Brenneman, G. Macdonald, and M. Roman. 2009. *Preschool pathways to science: Facilitating scientific ways of thinking, talking, doing, and understanding*. Baltimore, MD: Brookes.

Grady, J. 2010. The inquiry matrix: A tool for assessing and planning inquiry in biology and beyond. *The Science Teacher* 77 (8): 32–37.

Hoisington, C. 2010. Picturing what's possible—Portraits of science inquiry in early childhood classrooms. Paper presented at the SEED (STEM in Early Education and Development) conference, Cedar Falls, IA. *http://ecrp.uiuc.edu/beyond/seed/hoisington.html*.

Klentschy, M., and L. Thompson. 2008. *Scaffolding science inquiry through lesson design*. Portsmouth, NH: Heinemann.

Kutsunai, B., S. Gertz, and L. Hogue. 2003. *Squishy, squashy sponges*. Middletown, OH: Terrific Science Press.

Lehrer, R., and L. Schauble, eds. 2002. *Investigating real data in the classroom: Expanding children's understanding of math and science*. New York: Teachers College Press.

McGlashan, P., K. Gasser, P. Dow, D. Hartney, and B. Rogers. 2007. *Outdoor inquiries: Taking science investigations outside the classroom*. Portsmouth, NH: Heinemann.

National Academy of Engineering and National Research Council. 2014. *STEM integration in K–12 education: Status, prospects, and an agenda for research*. Washington, DC: National Academies Press.

National Science Foundation, Division of Elementary, Secondary, and Information Education (ESIE). n.d. *Inquiry: Thoughts, views, and strategies for the K–5 classroom.* Foundations monograph series, Vol. 2. Arlington, VA: National Science Foundation, ESIE. *www.nsf.gov/pubs/2000/nsf99148/start.htm.*

Ritz, W. C. 2007. *A head start on science: Encouraging a sense of wonder.* Arlington, VA: NSTA Press.

Rockwell, R. E., E. Sherwood, R. Williams, and D. Winnett. 2001. *Growing and changing.* White Plains, NY: Dale Seymour Publications.

Websites

American Society for Engineering Education. eGFI: Dream Up the Future [K-12 STEM education teacher resource]. *www.egfi-k12.org.*

Annenberg Learner. Learning science through inquiry [video series]. *www.learner.org/workshops/inquiry/videos.html?pop=yes&pid=1452.*

Boston Museum of Science. Engineering is elementary [curriculum products]. *www.mos.org/eie.*

Exploratorium. Institute for Inquiry: Examining the art of science education. *www.exploratorium.edu/education/ifi.*

First Hand Learning, Inc. Learning through direct experience. *www.firsthandlearning.org.*

Foundation for Family Science & Engineering. *www.familyscienceandengineering.org.*

National Association for the Education of Young Children. *Picturing good practice. You can count on math.* Handout 2: Math-related children's books, songs, and finger plays for preschoolers. *www.naeyc.org/files/tyc/file/BooksSongsandFingerPlays.pdf.*

National Association for the Education of Young Children (NAEYC) Early Childhood Science Interest Forum (ECSIF). On the NAEYC members-only interest forum page: *http://member-forums.naeyc.org;* on Facebook: *www.facebook.com/pages/Early-Childhood-Science-Interest-Forum-naeyc/140431919391071;* ECSIF blog: *http://ecsif.blogspot.com.*

PBS Parents. Science. *www.pbs.org/parents/education/science.*

STEMEd Hub. Science Learning through Engineering Design. *https://stemedhub.org/groups/sled.*

TryEngineering. *www.tryengineering.org.*

A Try Science conversation with Wynne Harlen [author of *The Teaching of Science in Primary Schools*]. *http://scienceonline.terc.edu/harlen_conversation/index.html.*

Appendix 2

National Science Teachers Association Early Childhood Science Education Position Statement

Introduction

At an early age, all children have the capacity and propensity to observe, explore, and discover the world around them (NRC 2012). These are basic abilities for science learning that can and should be encouraged and supported among children in the earliest years of their lives. The National Science Teachers Association (NSTA) affirms that learning science and engineering practices in the early years can foster children's curiosity and enjoyment in exploring the world around them and lay the foundation for a progression of science learning in K–12 settings and throughout their entire lives.

This statement focuses primarily on children from age 3 through preschool. NSTA recognizes, however, the importance of exploratory play and other forms of active engagement for younger children from birth to age 3 as they come to explore and understand the world around them. This document complements NSTA's position statement on elementary school science (NSTA 2002) that focuses on science learning from kindergarten until students enter middle or junior high.

Current research indicates that young children have the capacity for constructing conceptual learning and the ability to use the practices of reasoning and inquiry (NRC 2007, 2012). Many adults, including educators, tend to underestimate children's capacity to learn science core ideas and practices in the early years and fail to provide the opportunities and experiences for them to foster science skills and build conceptual understanding (NRC 2007, p. vii). Also underestimated is the length of time that young children are able to focus on science explorations. Effective science investigations can deeply engage young children for extended periods of time, beyond a single activity or session.

Source: National Science Teachers Association (NSTA). 2014. NSTA position statement: Early childhood science education. *www.nsta.org/about/positions/earlychildhood.aspx*.

NSTA supports the learning of science among young children that will create a seamless transition for learning in elementary school.

Young Children and Science Learning

NSTA identifies the following key principles to guide the learning of science among young children.

- Children have the capacity to engage in scientific practices and develop understanding at a conceptual level.

Current research shows that young children have the capacity for conceptual learning and the ability to use the skills of reasoning and inquiry as they investigate how the world works (NRC 2007, NRC 2012). For example, their play with blocks, water, and sand shares some science-relevant characteristics. Young children also can learn to organize and communicate what they learn, and know the difference between concrete and abstract ideas (Carey 1985). Adults who engage children in science inquiry through the process of asking questions, investigating, and constructing explanations can provide developmentally appropriate environments that take advantage of what children do as part of their everyday life prior to entering formal school settings (NAEYC 2013, p. 17; NRC 2007).These skills and abilities can provide helpful starting points for developing scientific reasoning (NRC 2007, p. 82).

- Adults play a central and important role in helping young children learn science.

Everyday life is rich with science experiences, but these experiences can best contribute to science learning when an adult prepares the environment for science exploration, focuses children's observations, and provides time to talk about what was done and seen (NAEYC 2013, p. 18). It is important that adults support children's play and also direct their attention, structure their experiences, support their learning attempts, and regulate the complexity and difficulty of levels of information (NRC 2007, p. 3). It's equally important for adults to look for signs from children and adjust the learning experiences to support their curiosity, learning, and understanding.

- Young children need multiple and varied opportunities to engage in science exploration and discovery (NAEYC 2013).

Young children develop science understanding best when given multiple opportunities to engage in science exploration and experiences through inquiry (Bosse, Jacobs, and Anderson 2009; Gelman, Brenneman, Macdonald, and Roman 2010).

The range of experiences gives them the basis for seeing patterns, forming theories, considering alternate explanations, and building their knowledge. For example, engaging with natural environments in an outdoor learning center can provide opportunities for children to examine and duplicate the habitats of animals and insects, explore how things move, investigate the flow of water, recognize different textures that exist, make predictions about things they see, and test their knowledge.

- Young children develop science skills and knowledge in both formal and informal settings.

Opportunities to explore, inquire, discover, and construct within the natural environment and with materials that are there need to be provided in formal education settings, such as preschool and early care and education programs through intentional lessons planned by knowledgeable adults. In addition, children need to have opportunities to engage in science learning in informal settings, such as at home with cooking activities and outdoor play or in the community exploring and observing the environment.

- Young children develop science skills and knowledge over time.

To effectively build science understanding, young children need opportunities for sustained engagement with materials and conversations that focus on the same set of ideas over weeks, months, and years (NRC 2007, p. 3). For example, investigating the concept of light and shadows over several weeks indoors and out with a variety of materials and multiple activities will allow children to re-visit and re-engage over time, building on observations and predictions from day to day.

- Young children develop science skills and learning by engaging in experiential learning.

Young children engage in science activities when an adult intentionally prepares the environment and the experiences to allow children to fully engage with materials. The activities allow children to question, explore, investigate, make meaning, and construct explanations and organize knowledge by manipulating materials.

Declarations

NSTA recommends that teachers and other education providers who support children's learning in any early childhood setting should

- recognize the value and importance of nurturing young children's curiosity and provide experiences in the early years that focus on the content and practices of science with an understanding of how these experiences connect to the science content defined in the *Next Generation Science Standards* (*NGSS*) (NGSS Lead States 2013);

- understand that science experiences are already a part of what young children encounter every day through play and interactions with others, but that teachers and other education providers need to provide a learning environment that encourages children to ask questions, plan investigations, and record and discuss findings;

- tap into, guide, and focus children's natural interests and abilities through carefully planned open-ended, inquiry-based explorations;

- provide numerous opportunities every day for young children to engage in science inquiry and learning by intentionally designing a rich, positive, and safe environment for exploration and discovery;

- emphasize the learning of science and engineering practices, including asking questions and defining problems; developing and using models; planning and carrying out investigations; analyzing and interpreting data; using mathematics and computational thinking; constructing explanations and designing solutions; engaging in argument from evidence; and obtaining, evaluating, and communicating information (NRC 2012, NGSS Lead States 2013);

- recognize that science provides a purposeful context for developing literacy skills and concepts, including speaking, listening, vocabulary development, and many others; and

- recognize that science provides a purposeful context for use of math skills and concepts.

NSTA recommends that teachers and other providers who support the learning of science in young children be given professional development experiences that

- engage them in learning science principles in an interactive, hands-on approach, enabling them to teach about science principles appropriately and knowledgeably;

- are ongoing and science-specific;

- help them understand how children learn science and engineering practices (NRC 2012, NGSS Lead States 2013);

- inform them about a range of strategies for teaching science effectively; and

- include the use of mentors to provide ongoing support for educators for the application of new learning.

NSTA recommends that those in a position to provide financial, policy, and other support for early childhood education should

- provide appropriate resources for teachers and children;

- ensure a positive and safe environment for exploration and discovery;

- ensure teachers receive sustained science-specific professional development that includes how children learn and how to teach science;

- provide mentoring; and

- establish a coherent system of science standards, instruction, appropriate assessment, and curriculum.

Parents and other caregivers can nurture children's natural curiosity about the world around them, creating a positive and safe environment at home for exploration and discovery. These recommendations can be found in NSTA's position statement, Parent Involvement in Science Learning (NSTA 2009), found at *www.nsta.org*.

—Adopted by the NSTA Board of Directors, January 2014

References

Bosse, S., G. Jacobs, and T. L. Anderson. 2009. Science in the air. *Young Children*, p. 10–15, reprinted and retrieved at *www.naeyc.org/files/yc/file/200911/BosseWeb1109.pdf*.

Carey, S. 1985. *Conceptual change in childhood.* Cambridge, MA: The MIT Press.

Gelman, R., K. Brenneman, G. Macdonald, and M. Roman. 2010. *Preschool pathways to science: Ways of doing, thinking, communicating and knowing about science.* Baltimore, MD: Brookes Publishing.

National Association for the Education of Young Children (NAEYC). 2013. All criteria document, 17–18. Retrieved from *www.naeyc.org/files/academy/file/AllCriteriaDocument.pdf*.

National Research Council (NRC). 2007. *Taking science to school: Learning and teaching science in grades K–8.* Washington, DC: National Academies Press.

National Research Council (NRC). 2012. *A framework for K–12 science education: Practices, crosscutting concepts, and core ideas.* Washington, DC: National Academies Press.

National Science Teachers Association (NSTA). 2002. NSTA Position Statement: Elementary School Science.

National Science Teachers Association (NSTA). 2009. NSTA Position Statement: Parent Involvement in Science Learning.

NGSS Lead States. 2013. *Next Generation Science Standards: For states, by states.* Washington, DC: National Academies Press.

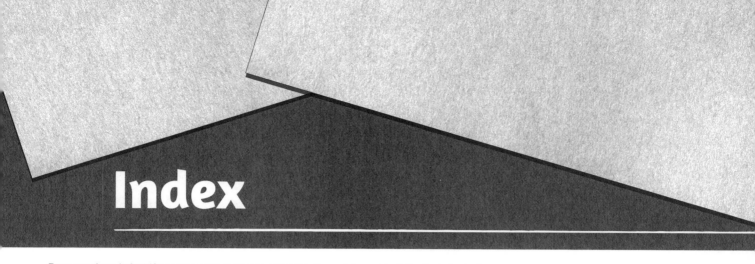

Index

Page numbers in **boldface** type refer to figures or tables; those followed by "n" refer to footnotes.

National Science Teachers Association